Synthesis Lectures on Engineering, Science, and Technology

The focus of this series is general topics, and applications about, and for, engineers and scientists on a wide array of applications, methods and advances. Most titles cover subjects such as professional development, education, and study skills, as well as basic introductory undergraduate material and other topics appropriate for a broader and less technical audience.

Horst Czichos

Introduction to Systems Thinking
and Interdisciplinary
Engineering

 Springer

Horst Czichos
University of Applied Sciences
BHT Berlin
Berlin, Germany

ISSN 2690-0300 ISSN 2690-0327 (electronic)
Synthesis Lectures on Engineering, Science, and Technology
ISBN 978-3-031-18241-9 ISBN 978-3-031-18239-6 (eBook)
https://doi.org/10.1007/978-3-031-18239-6

This Springer imprint is published by the registered company Springer Nature Switzerland AG
The registered company address is: Gewerbestrasse 11, 6330 Cham, Switzerland

Preface

The book describes technology and engineering with *Systems Thinking*. This is a way of exploring and describing complex entities by looking at connected wholes and relationships rather than separate parts.

I became familiar with the methodology of systems thinking—after studying engineering and physics—when I participated in an International Research Group of the Organization for Economic Co-operation and Development (IRG-OECD). In a Report (1974) it was concluded that the application of systems thinking and general systems theory is especially suited for developing a convenient framework for interdisciplinary engineering, bringing together its many aspects currently scattered through the scientific and technical literature.

Based on this experience, I applied systems thinking over the years in interdisciplinary research in Tribology at BAM, the German Federal Institute for Materials Research and Testing, and in my lectures on Mechatronics at the University of Applied Sciences, BHT Berlin.

The book presents the systems approach to technology in six chapters.

1. Scope of Technology
2. The System Concept
3. Tribological Systems
4. Mechatronic Systems
5. Cyber-Physical Systems and Industry 4.0
6. Systems Thinking in Health Technology.

I am grateful to Michael Luby, Executive Editor Springer Nature New York, who asked for a contribution to the new Springer Synthesis Lectures.

Berlin, Germany   Horst Czichos
June 2022

Contents

About the Author

Horst Czichos graduated in Precision Engineering and worked as design engineer in the optical industry. He holds degrees in Physics (FU Berlin) and in Materials Science (TU Berlin) and received from the University of Leuven an honorary doctorate for his interdisciplinary research in tribology. He was President of BAM (1992–2002), and of EUROLAB. Currently, he is Professor for Mechatronics at the University of Applied Sciences, BHT Berlin. In his publications, Dr. Czichos emphasizes the significance of systems thinking in science and technology. In 2013 he published a book on the triad Philosophy—Physics—Technology, which appeared as English Edition under the title *The Word is Triangular* (Springer 2021).

Scope of Technology

Technology today is marked by Friedrich Rapp in his book *Analytics for Understanding the Modern World* [1] as follows:

- *"Modern technology has emerged from the combination of craftsmanship and skill and scientific method. Generally speaking, technology is about objects and processes of the physical world, which are created by social action based on the division of labor. Products can be manufactured and procedures applied that were previously completely unknown and there is hardly any area of life that is not shaped by modern technology"* ....

The significance of technology for human life has been emphasized by the French philosopher Remi Brague in his book *The Wisdom of the World* [2]:

- *Technology is a kind of morality and maybe even the true morality. Today, technology is not only something that enables us to survive, it is more and more what enables us to live.*

The development of technology is closely linked to the progress of physics. Figure 1.1 compares their basic areas in a simplified overview. Mechanics is the basis for all mechanical technologies from mechanical engineering to civil engineering and aeronautics. Energy technology is based on thermodynamics. Electromagnetism is the physical basis for electrical engineering and the use of "energy from the socket" for all areas of human life. Electronics enables today's fast information and communication technology and, with computer networks, the global Internet. Optics opens up the universe through telescopes and the microcosm through microscopes. Quantum physics is the physical basis for micro-technology and nano-technology.

Technology is based on physics, but the methodology of technology is fundamentally different from the methodology of physics.

© The Author(s), under exclusive license to Springer Nature Switzerland AG 2022   1
H. Czichos, *Introduction to Systems Thinking and Interdisciplinary Engineering*,
Synthesis Lectures on Engineering, Science, and Technology,
https://doi.org/10.1007/978-3-031-18239-6_1

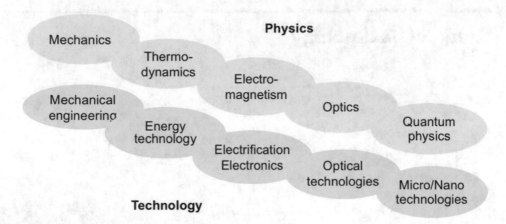

Fig. 1.1 Physics and technology in a comparative presentation of their fundamental areas

- Physics explores nature with the methods of "reductionism and analysis", founded by Descartes and Newton: breaking down a problem into its simplest elements and their analysis for the recognition of elementary forces and natural constants.
- In contrast, technology requires a "holistic" methodology for the development, production and application of products and technical systems. Fundamental to technology are systems thinking and interdisciplinary engineering.

1.1   Basics of Technology and Engineering

Engineering sciences are the sciences that deal with the research, development, design. production, and application of technical products and technical systems. The classical engineering sciences are civil engineering, mechanical engineering, electrical engineering and production engineering. More recent courses of study are precision engineering, energy technology, chemical engineering, process engineering, environmental technology and technical computer science. In the twentieth century, new interdisciplinary courses of study were created in line with the growing needs of technology, business and society, e.g. mechatronics as a combination of m*echanics*—electronics—computer science or industrial engineering by combining economics (business administration, economics) and law with one or more engineering sciences to form a separate field of knowledge.

Based on the state of science and technology in the twenty-first century and the curricula of technical universities and colleges, the fundamentals of engineering sciences can be presented in a "knowledge circle" with four basic areas of disciplines, see Fig. 1.2:

1. Mathematical and scientific fundamentals Mathematics, Statistics, Physics, Chemistry

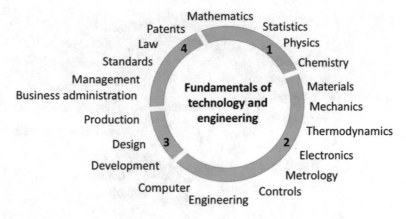

Fig. 1.2   The basic subjects of engineering sciences represented as "knowledge circle"

2. Technological fundamentals Materials, Mechanics, Thermodynamics, Electronics, Metrology, Controls, Computer Engineering
3. Fundamentals for products and services Development, design, production
4. Economic and legal fundamentals Business administration, management, standardization, law, patents.

The individual disciplines required for interdisciplinary engineering sciences can be assembled from the knowledge circle in a modular fashion. For example, the collection of fundamentals for the curriculum of *Industrial Engineering* is illustrated Fig. 1.3 The collection of fundamentals for the curriculum of *Mechatronics* is illustrated Fig. 1.4.

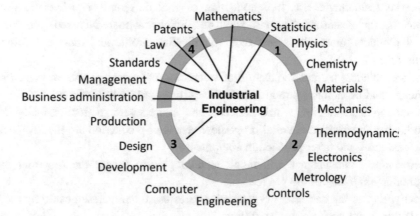

Fig. 1.3   Main disciplines for Industrial Engineering, selected from the circle of knowledge

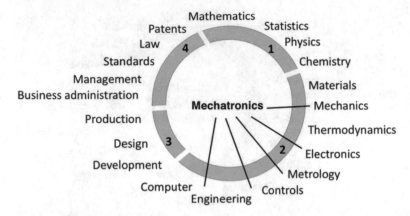

Fig. 1.4 Main disciplines for Mechatronics, selected from the knowledge circle

1.2 The Product Cycle

The technology product cycle illustrates that all technical products—accompanied by the necessary flow of materials, energy, information, manpower and capital—move in "cycles" through the techno-economic system. From resources of matter and raw materials, the engineering materials—metals, polymers, ceramics, composites—are processed, and transformed by design and manufacture into products and systems. At the end of use, deposition or recycling of scrap and waste is necessary, Fig. 1.5. The production cycle requires the succession of several technologies:

- Raw material technologies for the exploitation of natural resources.
- Material technologies for the production of materials and semi-finished products from the raw materials. The structural and functional materials, required for technical products, must correspond to the application profile and must accordingly be optimized.
- Design methods, i.e. the application of scientific, mathematical and art principles for efficient and economical structures, machines, processes, and systems.
- Production technologies, which concerns the manufacture of geometrically defined products with specified material properties. Means of production are plants, machines, devices, tools and other production equipment.
- Performance, maintenance and repair technologies to ensure the functionality and economic efficiency of the product.
- Finally, processing and recovery technologies to close the material cycle by recycling or, if this is not possible, by landfilling.

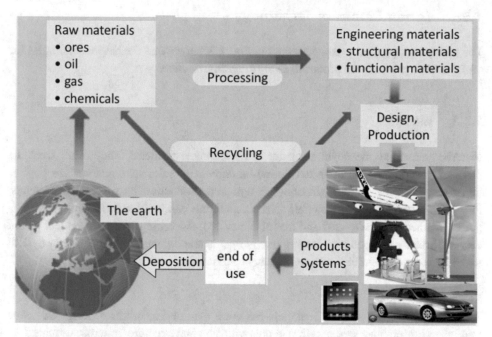

Fig. 1.5 Illustration of the technology product cycle

From an economic point of view, the production cycle can be regarded as an "economic value-added chain".

The illustration of the production cycle, Fig. 1.5, reminds us that the components of technical products and systems interact with their environment in their technical function. These interactions are generally described as two complementary processes:

- Immission, the exposure of substances or radiation on materials and technical products, which can lead to corrosion, for example.
- Emission, the discharge of substances (e. g. CO_2) or radiation (also sound). An emission from a material is usually at the same time an immission into the environment.

To protect the environment—and thus people—there are legal regulations for emission and immission control with procedural regulations and limits for harmful substances and radiation. With regard to the protection of the environment, the main requirements are:

- Environmental compatibility, the property of not adversely affecting the environment in its technical function (and on the other hand not being adversely affected by the environment in question).
- Recyclability, the possibility of recovery and reprocessing after the intended use.
- Waste disposal, the possibility of disposing of material if recycling is not possible.

1.3 Corner Stones of Technology

The technology product cycle, presented in Fig. 1.5, shows that the base technologies for all fabricated products are *Material, Energy and Information.*

1.3.1 Material

Materials constitute the physical matter of all technical products. They result from the processing and synthesis of matter, based on chemistry, solid state and surface physics transformed by materials engineering, design and production into technical products, Fig. 1.6. From the view of materials science, materials can be of natural (biological) origin or synthetically processed and manufactured [3]. According to their chemical nature they are broadly grouped traditionally into inorganic and organic materials. Their physical structure can be crystalline, or amorphous, or mixtures of both structures.

Scale of Materials
The geometric length scale of materials has more than twelve orders of magnitude. The scale spectrum embraces nanoscale materials with quantum structures, microscale materials, and materials with macroscale architectures for macro engineering, assembled structures and engineered systems. Figure 1.7 illustrates the dimensional scale relevant for today's technology.

Processing of Materials
For their applications, materials have to be *engineered* by processing and manufacture in order to fulfil their purpose as constituents of products, designed for the needs of economy and society. There are the following main technologies to transform matter into engineered materials, see Fig. 1.8:

- Net forming of suitable matter (liquids, molds), e.g. LASER printing

Materials originate from synthesis and processing of matter. They can be of natural (biological) or synthetic origin and are transformed by design and production into engineered products.

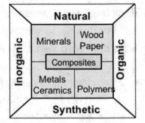

Matter $\xrightarrow[\text{Processing}]{\text{Synthesis}}$ Material $\xrightarrow[\text{Production}]{\text{Design}}$ Product

Fig. 1.6 Genesis of products and types of materials

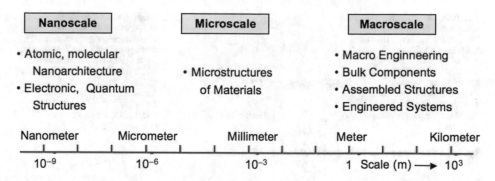

Fig. 1.7 Scale of material dimensions

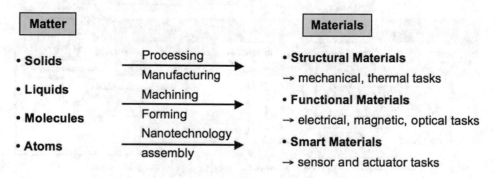

Fig. 1.8 Materials and their characteristics result from the processing of matter

- Machining of solids, i.e. shaping, cutting, drilling, etc.,
- Nano-technological assembly of atoms or molecules.

Characteristics of Materials

Experience shows that materials have four basic characteristics, depicted as tetrahedron to illustrate that they are interrelated [4], Fig. 1.9.

In a general compilation, the basic scientific, engineering and technological aspects of materials can be jointly depicted in a holistic view, Fig. 1.10.

The properties of materials, which are of fundamental importance for their engineering applications, can be categorized in three basic groups, broadly classified in the following categories:

- **Structural materials**: engineered materials which have specific mechanical or thermal properties in responding to an external loading by a mechanical or thermal action.

Composition and microstructure of materials are *intrinsic (inherent)* characteristics. They result from the processing and synthesis of matter and characterize the chemical nature and physical stricture of a material. Composition and microstructure of materials are determined by chemical and microstructural analysis.

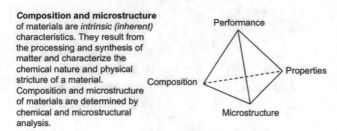

Properties and performance of materials are *extrinsic (procedural) characteristics*. They describe the response of materials to functional loads (stress) and environmental influences. They are also influenced by design and production. Properties and performance of materials are determined by materials testing and measurements.

Fig. 1.9 The basic interlinked characteristics of materials

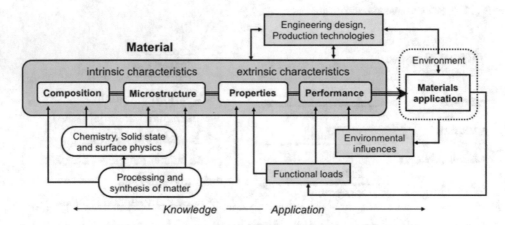

Fig. 1.10 Overview of the scientific, technological and engineering aspects of materials

- **Functional materials**: engineered materials which have specific electrical, magnetic or optical properties in responding to an external loading by an electromagnetic or an optical action.
- **Smart materials** :engineered materials with intrinsic or embedded *sensor* and *actuator functions*, which are able to accommodate materials in response to external loading, with the aim of optimizing material behavior according to given requirements for materials performance.

Numerical values for the various materials properties can vary over several orders of magnitude for the different material types. An overview on the broad numerical spectra of some mechanical, electrical and thermal properties of metals, inorganics and organics are shown in Fig. 1.11.

It must be emphasized that the numerical ranking of materials in Fig. 1.11 is based on "rough average values" only. Precise data of materials properties require the specification of various influencing factors described above and symbolically expressed as follows:

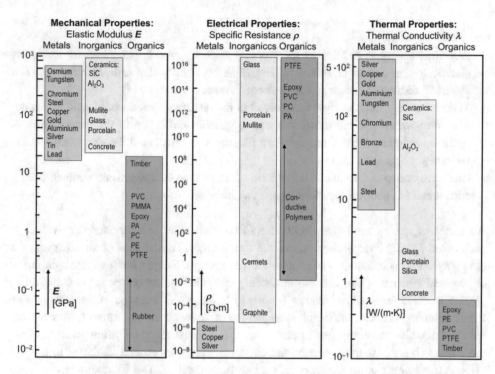

Fig. 1.11 Overview of mechanical, electrical, and thermal materials properties for the basic types of materials (metal, inorganic, or organic)

Materials properties data = f (composition-microstructure-scale; external loading; …).

1.3.2   Energy

Energy is one of the fundamentals areas of physics and is divided into the following categories:

- Mechanical energy
  - Potential mechanical energy characterizes the working capacity of a physical system, which is determined by its position in a force field.
  - Kinetic mechanical energy is the work capacity that an object can generate due to of his movement. It corresponds to the work that must be done to move the object

from rest to momentary motion. It depends on the mass and the velocity of the moving body.

- Electrical energy is defined as the working capacity that is generated by means of electricity, transmitted or stored in electric fields. Energy that is transferred between electrical energy and other forms of energy means electrical work.
- Thermal energy (heat energy) is involved in the disordered motin of atomic or molecular components of a substance. It is a state variable and is part of the inner energy. A radiation field has thermal energy when its energy is distributed disorderly among the various possible waveforms.
- Chemical energy is distributed in the form of a chemical compound, stored in an energy carrier and in chemical reactions which can be released.

All forms of energy are largely convertible into each other and therefore equivalent to each other. During energy conversions, the total energy, i.e. the sum of all energies of a closed system is constant (law of conservation of energy). **Energy is the working capacity of physical systems.** Thus, it is of outstanding technical and economic importance.

The **metrological unit of energy** (or work) is the joule. Because of the diversity of forms of energy and the different value ranges of released energy, different units have developed, which are equivalent to each other. Mechanical energy is often given in Newton meters (N · m), thermal energy and chemical energy in general in the SI unit joule (J), electrical and magnetic energy in watt seconds (W · s). The following equivalence applies: $1\,J = 1\,N \cdot m = 1\,W\,s$.

Energy technology is an interdisciplinary engineering science, with the technologies for the production, conversion, storage and use of energy. Primary energy production refers to the first stage where energy enters the supply chain before any further conversion or transformation process. Energy carriers are usually classified as

- Fossil, using coal, crude oil and natural gas,
- nuclear, using uranium,
- Renewable, using hydro power, wind and solar energy, biomass, among others.

The energy content of energy sources is either heat that can be used for household and industrial purposes, or the energy content is made available by mechanical, thermal, or chemical transformation as electrical energy (electricity production). In a power station, mechanical energy is usually generated by means of generators into electrical energy, which is usually fed into the power grid. Mechanical (kinetic) energy for drive of generators comes from water or wind movements or uses—via steam turbines or gas turbines—thermal energy from burning coal, oil, natural gas, biomass, garbage or nuclear energy. Photovoltaic systems convert radiation energy directly into electricity. Important framework conditions for energy technology are

(a) The availability of energy sources,
(b) Environmental aspects (e.g. pollutant emissions),
(c) Energy conversion efficiency, the ratio of useful energy to primary energy, expressed as efficiency $\eta = E_{output}/E_{input}$ in percent.

Examples of energy conversion efficiency:

- Heat production:
  - Gas heating 80 … 90%
  - Coal-fired furnace (industry) 80 … 90%
  - Solar collector up to 85
  - Electric range (household) 50 … 60%.
- Power generation:
  - Hydropower plant 80 … 90%
  - Gas/steam turbine power plant (natural gas) 50 … 60
  - Coal-fired power station 25 … 50%
  - Wind turbine up to 50
  - Nuclear power plant 33%
  - Solar cell 5 … 25.

Energy is an indispensable resource for human existence [5]. Just to stay alive, a human being needs an average daily food intake of about 3000 kilocalories (kcal), which represents a food-related energy equivalent of about 1200 kilowatt hours (kWh) per year. The total consumption requirement of people to energy (heating, electricity, fuels, etc.) has been estimated to be about 15 times higher. With a world population of more 7 billion people, the annual global energy demand sums up to about 140,000 billion kWh (500 × 10^{18} J) primary energy.

An overview of energy technology with regard to the energy end users is given in Fig. 1.12. In Germany, the primary energy demand accounts for approximately 2.8% of the world's energy demand [6]. The basic feature of electricity energy is shown in Fig. 1.13.

Electrical energy is the most versatile energy source, which can be converted into other forms of energy with very low losses. The availability of electric power is a basic prerequisite for any modern industry and cannot be replaced by other energy sources. Experience shows that a failure of the electricity supply brings any economy to a standstill and must therefore be limited as far as possible. A high level of security of supply of electricity is therefore an important condition for technology, economy and society.

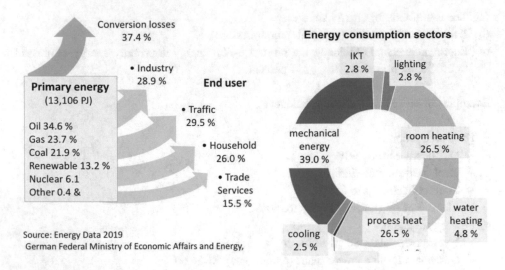

Fig. 1.12 Overview of energy technology: primary energy, end user and energy consumption sector (note that the figures are country-specific and year-dependent)

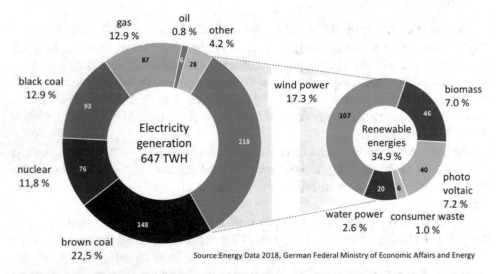

Source: Energy Data 2018, German Federal Ministry of Economic Affairs and Energy

Fig. 1.13 The basic feature of electricity energy technology (note that the figures are country-specific and year-dependent)

1.3.3   Information and Digitalization

Information is the third basic technology besides energy and material. The umbrella term *information and communications technology (ICT)* refers to the methods and processes of generating, transmitting, storing and applying information (data, numbers, charac-ters) for use by science, technology, business and society. The scientific foundation of today's information and communication technology is *digitalization* (or digitization) *of information,* mathematically based in the "Binary System", Fig. 1.14.

Digitalization is the conversion of text, pictures, or sound into a digital form that can be processed by a computer. The principle is illustrated in Fig. 1.15. An information, acquired by a sensor (sensor input) is first converted in an analog electrical signal (sensor output) and then converted into a binary signal sequence by an analog-to-digital AD converter. Since the processing of digital (binary) signals only requires a distinction between two signal states (0 or 1 or low or high), error-free information transfer to a computer and data storage is possible in principle. The information is usually transmitted in cryptographically secure information data blocks (block chains). Figure 1.15 provides an overview of the methodology of digitalization for information and communication technology.

Fig. 1.14 The binary system is the mathematical base of digital technologies

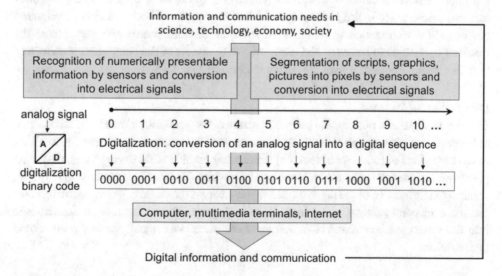

Fig. 1.15 The methodology of digitalization

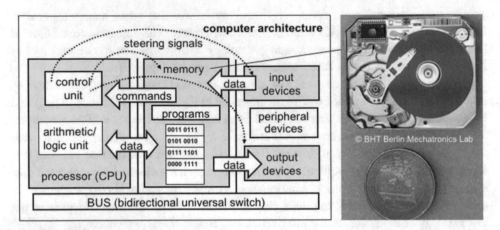

Fig. 1.16 The computer architecture illustrated by a block diagram and a photo of a hard disc drive

In the digital world of information and communication, **Computers** have a central role in information and communication technology. They are technical systems that can be instructed by an operator to carry out arithmetic or logical operations automatically via computer programming. Modern computers have the ability to follow generalized sets of operations, called programs. These programs enable computers to perform an extremely wide range of tasks.

A computer consists of a central processing unit (CPU), and a memory, typically a hard disc drive (HDD). The processing element carries out arithmetic and logical operations. Peripheral devices include input devices (keyboards, mice, etc.), output devices (monitor screens, printers, etc.), and input/output devices that perform both functions. Peripheral devices allow information to be retrieved from an external source and they enable the result of operations to be saved and retrieved. An overview of the computer's architecture together with a photo of a hard disc drive is given in Fig. 1.16.

Digitization of Pictures
In order to digitize a picture, the picture is scanned, i.e. split into a matrix with rows and columns (the picture is "pixelated"). This can be done by scanners, photography, satellite-based or medical sensors. For a black and white raster graphic without grey tones, the image pixels are assigned the values 0 for black and 1 for white. The matrix is read out line by line, resulting in a sequence of the digits 0 and 1 that represents the picture. When digitizing color images, each color value of a pixel in the RGB (red–green–blue) color space is decomposed into the values red, green and blue, and these are stored individually with the associated quantization.

Text Digitization

When digitizing text, the document is first digitized in the same way as an image, i.e. scanned. If the original appearance of the document is to be reproduced as accurately as possible, no further processing takes place and only the image of the text is stored. If the linguistic content of the documents is of interest, the digitized text image is translated back into a character set by a text recognition program, e.g. using the ASCII code (American Standard Code for Information Interchange) or, in the case of non-Latin characters, the Unicode (ISO 10646). Only the recognized text is then saved.

Digitization of Audio/Video Signals

The digitization of audio data is called sampling. Sound waves are converted into analog electronic vibrations with a microphone as sensor and measured and stored as digital values. These values can also be "put together" to form an analog sound wave, which then can be made audible. Due to the large amounts of data that are generated, compression methods are used. This allow the realization of more space-saving data carriers, e.g. FLAC (Free Lossless Audio Codec) or MP3 (lossy compression of digitally stored audio data).

Optical data memories store data sequences in the form of pits and lands (plateau) in a compact disc (CD) for audio data or in a digital video disc (DVD) for video data.

The data track information is read out in a CD/DVD player by mechanical contactless scanning with a LASER scanning unit when the CD/DVD is rotated. Pits and lands have a height difference leading to LASER-beam interference. The alternation of pits and lands is detected as a dark/light change (bit "1"). The laser beam reflection at individual pits or lands returns bits "0". This results in a serial 1–0 data stream (data output), which is fed to a photo detector, giving an output as an analogue audio or video signal sequence will.

1.4   Greatest Engineering Achievements

A comprehensive analysis of the situation and significance of technology at end of the twentieth century was conducted by the US National Academy of Engineering (www. nae.edu). The Greatest Engineering Achievements are shown in Fig. 1.17 in the order specified by the Academy.

The most important technology is ELECTRIFICATION. Its worldwide importance for all areas of human life as well as for all of technology, economy and society is obvious, and is clearly illustrated by the scenario of the catastrophic consequences of blackout: collapse of information and communication possibilities (telephone, TV, Internet) and failure of the entire electrified infrastructure, from rail transport to drinking water supply. Because of its central importance for technology, economy and society, ELECTRIFICA-TION is called the "Workhorse of the Modern World" by the Academy. The other major

1. Electrification	11. Highways
2. Automobile	12. Spacecraft
3. Airplane	13. Internet
4. Water Supply and Distribution	14. Imaging
5. Electronics	15. Household Appliances
6. Radio and Television	16. Health Technologies
7. Agricultural Mechanization	17. Petrochemical Technologies
8. Computers	18. Lasers and Fiber Optics
9. Telephone	19. Nuclear Technologies
10. Air Conditioning and Refrigeration	20. High-performance Materials

Fig. 1.17 The greatest engineering achievements

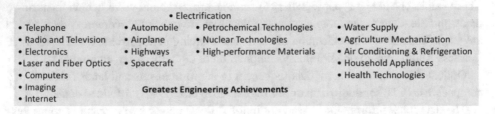

Fig. 1.18 The greatest engineering achievements, displayed in essential technology areas

engineering achievements are disposed in four essential technology groups for human wants in Fig. 1.18.

- The first group concerns the large and complex field of information and communication such as telephone, radio, television, electronics, computers and the Internet.
- The second group names the most important technologies for human mobility, the automobile and the airplane.
- The third group concerns "enabling technologies", such as petrochemicals and high-performance materials.
- The fourth group includes technologies that are vital to the world's population, from water supply technology and agricultural engineering to health technology.

The overview of the "greatest engineering achievement" at end of the twentieth century and their grouping according to human needs—from communication and human mobility to water supply and health technologies—underline the indispensable role of modern technology for industry, economy, society—and human life in general!

New developments—especially the "digitalization of information" and the "Internet of Things"—emerges in recent years. The **scope of today's technology** can be characterized as follows:

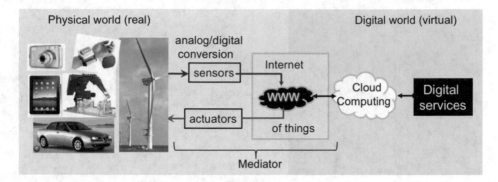

Fig. 1.19 Technology in the twenty-first century

Technology in the twenty-first century is a combination of the "real physical world" of technical objects with the "virtual digital world" of information and communication [7]. In addition, a supportive "Mediator" between the two worlds is essential for advanced technology as illustrated in Fig. 1.19.

References

1. Rapp, F.: Analytics for Understanding the Modern World. Alber, Freiburg (2012) (in German)
2. Brague, R.: The Wisdom of the World. Beck, München (2006) (in German)
3. Hummel, R.E.: Understanding Materials Science. Springer, New York (2004)
4. Olson, G.B.: Designing a new material world. Science **288**, 993–998 (2000)
5. Heinloth, K.: Energy for our life. In: Martinssen, W., Röß, D. (eds.) Physik im 21. Jahrhundert. Springer, Berlin (2011) (in German)
6. German Federal Ministry of Economic Affairs. Energy Data (2019)
7. Czichos, H.: The World is Triangular. Springer, Berlin (2021)

The System Concept

2

The classical method of scientifically analyzing objects of interest is "analytical reductionism". An entity, i.e. the object of an investigation, is virtually broken down into its individual parts so that each part could be analyzed and described. This principle can be applied analytically in a variety of directions, e.g. resolution of causal relations into separate parts, searching for "atomic units" in science or for "material constants" in engineering. Application of the classical analytical procedure depends on the condition that interactions between the Parts are non-existent or, at least, weak enough to be neglected for certain research purposes.

The systems approach goes beyond the methodology of analytical reductionism. It considers how the parts interact with the other constituents of the system forming an entity of "organized complexity" [1].

2.1 General Systems Theory

In this approach, a system is a set of elements interconnected by structure and function. The behavior of a system is the manner in which the whole or parts of a system act and react to perform a function. In characterizing the behavior of systems, the terms structure and function also must not be isolated from each other, because the structure and the function of systems are interconnected.

- System definition: A system is a set of elements interconnected by structure and function

    (I) Structure: $S = \{A, P, R\}$
        A Elements (components)

© The Author(s), under exclusive license to Springer Nature Switzerland AG 2022
H. Czichos, *Introduction to Systems Thinking and Interdisciplinary Engineering*,
Synthesis Lectures on Engineering, Science, and Technology,
https://doi.org/10.1007/978-3-031-18239-6_2

$$A = \{a_1, a_2, \ldots, a_n\} \quad \text{n: number of elements}$$

P Properties of the elements

$$P = \{P(a_i)\} \quad i = 1 \ldots n$$

R Relations (interactions) between elements

$$\{R(a_i \leftrightarrow a_j)\} \quad i, j = 1 \ldots n, \, j \neq i$$

(II) Inputs (X), Outputs (Y)

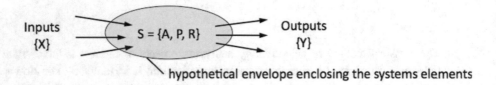

hypothetical envelope enclosing the systems elements

(III) Function
     − Support of loads
     − Transfer or transformation of operating Inputs into functional outputs

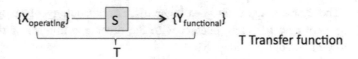

T Transfer function

- System behavior: the manner in which a system acts and reacts in performing a function.

The abstract formalism of the General Systems Theory should be supplements by a classification scheme indicating the "type" of system under study. According to Norbert Wiener, the founder of Cybernetics [2], the input and output of systems may be broadly classified into the three categories *Material, Energy* and *Information*. As physical or engineering systems are composed of materials components, whose properties, interactions, etc. may change with time, the parameter *time* is an independent attribute of any of these systems.

2.2 Technical Systems

For the characterization of multicomponent technical products, instead of poorly distinguishable expressions such as "machine", "device", "apparatus", the generic term "Technical System" has been fixed.

- Technical Systems are characterized by their function to generate, convert, transport and/or store material, energy and/or information. They have a spatial structure and are composed of components, which emerge from design and production technology (Brockhaus Encyclopedia, 2000).
- Technical Systems are designed and build by Interdisciplinary Engineering.

2.2.1 Mechanical and Electrical 2-Body Systems

The application of the system concept to the characterization of technical devices is exemplified for the simplest case of "2-body systems", namely a mechanical gear pair and an electrical transformer, in Fig. 2.1. The **function** of both systems is to transfer operating inputs, namely.

System characteristics	Mechanical 2-body system: gear pair	Electrical 2-body system: transformer
System pictogram		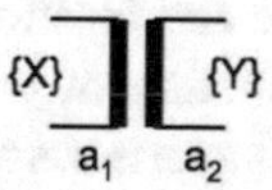
Structure A Elements	a_1 driving wheel a_2 driven wheel	a_1 primary coil a_2 secondary coil
P Properties $P(a_1)$, $P(a_2)$	gear type (e. g. spur, helical. hypoid) pitch diameters, modules, etc	coil type (e. g. single, bifilar) number of turns, etc.
R Relations $R(a_1 / a_2)$	Hertzian tribo-contcact mechanics traction, friction, pitting, etc.	inductive coupling eddy current effects, etc.
Function X Inputs	input speed n_x input torque τ_x	input voltage u_x input current i_x
Y Outputs	output speed n_y output torque τ_y	output voltage u_y output current i_y
T Transfer function	$T: (n_x, \tau_x) \longrightarrow (y_y, \tau_y)$	$T: (u_x, i_x) \longrightarrow (u_y, i_y)$

Fig. 2.1 The characteristics of technical 2-body systems

- speed and torque in the mechanical system, and
- voltage and current in the electrical system

into transformed functional outputs for the intended technical purpose.

Regarding the **structure** of the two systems, a fundamental difference between the electrical and the mechanical system has to be noted with respect to their performance:

- In the electrical system, the function of the system is realized through electromagnetic coupling processes between the two elements of the system. The macroscopic structure of the electrical 2-body system remains constant with time. A possible submicroscopic deterioration process can be electromigration.
- In the mechanical system, the function is realized through (Hertzian) contact mechanics and frictional traction of the interacting gears, inevitably connected with friction-induced energy losses. The interfacial functional stress may cause geometrical changes of the gear pair ("pitting" type of wear). Because the mechanical system operates with "interacting surfaces in relative motion", it is a "tribological" system.

2.2.2 Multicomponent Technical Systems

Emanating from the principles of General Systems Theory and engineering experience, the basic features of Multicomponent Technical Systems can be described as follows:

- A technical system is a set of elements (engineered components) interconnected by structure and function.
- The **Function** of technical systems is to generate, store, transmit or/and transform *energy*, *materials* and/or *information*.
- The **Inputs** are to be distinguished as
  - operating inputs $\{X\}$, responsible for the functional task of the system, Note: inputs can execute "loads" and "stress" on the structural components of a system
  - auxiliary inputs $\{X_A\}$, (e.g. energy support),
  - disturbances $\{Z\}$ from the environment (e.g. vibrations, moisture, electrical smog).
- The **Outputs** are to be distinguished as
  - functional outputs $\{Y\}$, responsible for the functional task of the system
  - Loss outputs $\{Y_L\}$, can be
        energy loss, (e.g. due to friction),
        material loss or emissions, (e.g. wear debris, CO_2 emission),
        information loss, (e.g. analog/digital (A/D) mistakes).
- The **Structure** $S = \{A, P, R\}$ of technical systems consists of multiple components $\{A\}$, designed and manufactured with their properties $\{P\}$ and interactions $\{R\}$ for a given task. If the structure of a system changes with time t, for example due to

detrimental changes of element properties (P) or relations (R) under working loads L (e.g. force, temperature), the structure of a system is load-and-time dependent and is represented by the set

$$S(t, L) = \{A, P(t, L), R(t, L)\}$$

The basic features of Technical Systems can be illustrated in a "holistic view", Fig. 2.2.

There is a broad spectrum of multicomponent technical systems. With regard to their functional areas and structural features, the following main categories of technical systems can be distinguished:

- Technical systems for the realization of motion and the transmission of mechanical energy with *interacting surfaces in relative motion*, especially in mechanical engineering, production, and transport.
  - Tribological Systems
- Technical systems that require for their tasks the combination of mechanics, electronics, controls, and computer engineering.
  - Mechatronic Systems
- Technical systems with a combination of mechatronics and internet communication.
  - Cyber-physical Systems (CPS)

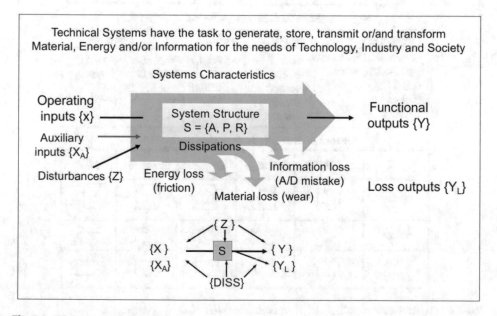

Fig. 2.2 Holistic view of technical systems

- Technical systems for the digitalization of technology and engineering supported by CPS.
  - Industry 4.0
- Technical systems developed to solve a health problem and improve quality of lives.
  - Health Technology.

2.3   Physical State Variables and Process Elements

A starting point for the consideration of physical state variables of technical systems is the "Quadrupole Model". In this model, originating from electrical engineering, the operating input parameters of technical systems are the variables *effort* ε and *flow* ϕ. They are **State Variables** of technical systems and their product gives the physical power $p = \varepsilon \cdot \phi$. The state variables for different system types are listed in Fig. 2.3.

Process Elements for the generation and processing of the process variables in technical systems are described with key words in the lower part of Fig. 2.3.

Quadrupole Model	flow $\phi_{in} \longrightarrow \phi_{out}$ effort ε_{in} Element ε_{out}	Physical State Variables ε : effort; ϕ : flow power $p = \varepsilon \times \phi = dE/dt$ E: energy flow

Physical State Variables in Technical Systems

system type	effort	flow	power
• mechanical, translation	force F	velocity $v = dx/dt$	$F \cdot v$
• mechanical, rotation	torque T	angular velocity ω	$T \cdot \omega$
• electrical	voltage V	current I	$V \cdot L$
• fluidic	pressure p	volume flow $\dot{V}$	$p \cdot \dot{V}$
• thermal	temperature τ	heat flow $\dot{Q}$	$\tau \cdot \dot{Q}$

Process Elements in Technical Systems

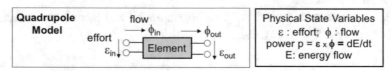

	Source	Storage	Transformer	Converter	Sink
symbol					
effect	output from stock	fill-in and take-out	input-output linkage	input-output change	input dissipation
ϕ flow; ε effort					

Fig. 2.3   Physical state variables and process elements in technical system types

They are "building blocks" of technical systems. A prominent example which is designed with all process elements shown in Fig. 2.3 is a Wind Energy Converter—the most important technical system for renewable energy technology—see Fig. 4.30.

The functional behavior of a technical system can be categorized in different states, as illustrated in Fig. 2.4.

Stationary behavior

If the inputs and outputs are stationary (time-independent) and the structure of the system $S = \{A, P, R\}$ is stable, the functional outputs $\{Y\}$ may be describable as functions of the operating inputs $\{X\}$ as linear input–output function.

Dynamic behavior

If the inputs and outputs vary with time, the system is said to be in a "dynamic state". The dynamic (time-dependent) behavior of a system is conventionally characterized by two different input test functions:

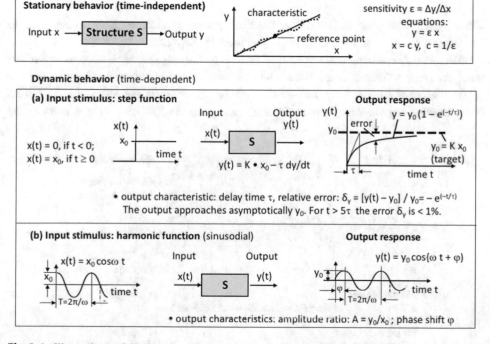

Fig. 2.4 Illustrations of the functional behavior of technical system

(a) a harmonic undamped sinusoidal input x gives a sinusoidal output y with an amplitude ratio y_0/x_0 and a phase shift of the output signal. Such behavior is called in the terminology of control theory "second-order behavior".

(b) a step function input x gives an output response y in the form of a first-order differential equation. This is called in the terminology of control theory a "first-order behavior". It is characterized by a delay time constant τ which is a measure of the system's inertia.

Stochastic behavior

In some situations, the functional input–output relations may be influenced by stochastic processes, i.e. dynamic effects of uncertainty and random disturbances ("noise"). In such cases, an estimate of the limits of proper systems behavior by means of the theory of probabilities may be attempted.

2.4   Modeling of Technical Systems

The physical background of modeling physical systems is given by Richard Feynman in his famous *Lectures on Physics* [3] and illustrated in Fig. 2.5:

*We can study electrical and mechanical systems by the **principle that the same equations have the same solutions.** In an electrical circuit with a coil of inductance L and a voltage V, the electrical current I depends on the voltage according to the relation $V = LdI/dt$. This equation has the same form as Newton's law of motion. The rate of doing work on the*

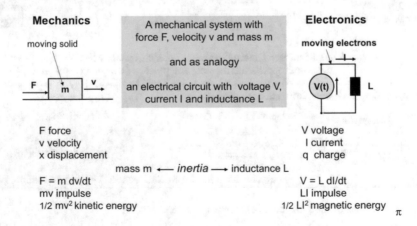

Fig. 2.5 Analogy between simple mechanical and electrical systems

inductance is voltage times current, and in the mechanical system it is force times velocity. Therefore, in the case of energy, the equations not only correspond mathematically, but also have the same physical meaning as well.

In the modeling of technical systems, often simplified "modules" are considered, consisting in the mechanical case of a mass-spring-damper structure and in the electrical case of a inductance-resistance–capacitance structure. This modeling approach is illustrated in Fig. 2.6.

The analogy between mechanical and electrical modules can be used for the computer-aided modeling of technical systems. An example is illustrated in Fig. 2.7.

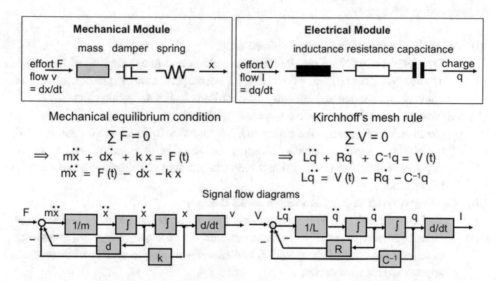

Fig. 2.6 Model representations of technical systems

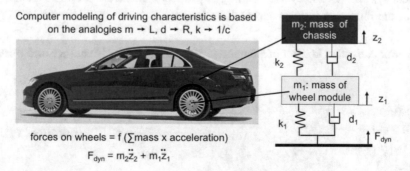

Fig. 2.7 An example of modeling

2.5   Design of Technical Systems

General *Principles of Engineering Design* have been suggested by the Royal Academy of Engineering (www.raeng.org.uk) at the end of the twentieth century with the following statement:

"In engineering it is possible to identify basic principles which can be referred to by all designers of any discipline. These principles are intended to provide a total context for good design. They are not necessarily rooted in physics or mathematics, and derive more from experience, practice or pragmatism than from formal theory. They are the substance of professional engineering judgement. It is possible to consider engineering design as encompassing three stages:

(i)   All design begins with a clearly defined need.
      The definition of Need requires the recognition and understanding of the nature of society, of economics, of humanity's needs. The human qualities of reason, compassion, service and curiosity all contribute to the definition of need.
(ii) All designs arise from a creative response to a need.
      Creative Vision requires the ability to think laterally, to anticipate the unexpected, to delight in problem solving, to enjoy the beauties of mind as well as of the physical world. The ethos within which the problem is being addressed must be understood.
(iii) All designs result in a system which meets the need.
      To Deliver a solution to the recognized need requires the assembly and management of resources and of team members with the necessary skills and knowledge of natural laws and of the materials and energies needed to affect an efficient and appropriate creative design".

This very general design principles can be concretized by combining the description of technical systems (Fig. 2.2) with the design principle "Structure follows Function". This leads to the following systems approach to design:

Definition: A System is a Set of Elements Interconnected by Structure and Function

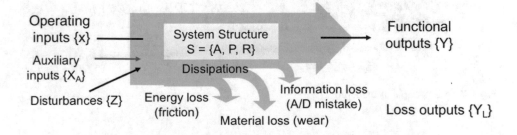

Design Principle: Structure Follows Function

1. Define the function of the system and specify the necessary input and outputs; consider also auxiliary inputs and disturbances.
2. Realize the structure of the system with engineered components {A}, designed and manufactured with their properties {P} and interactions {R} to fulfil the function; consider also functional stress on the structural elements and dissipations.

In addition to the interdisciplinary design of the system it may be necessary to

- install a control module with sensors, actuators and appropriate algorithms,
- monitor the system with structural health monitoring and performance control.

2.6 Performance and Reliability of Technical Systems

In their applications, technical systems have to fulfil various tasks. The material components have to bear "functional stress" of mechanical, thermal and electromagnetic nature. They are also in contact with other solid bodies, aggressive gases, liquids or biological organisms because there is no usage of materials without interaction with the environment [4]. All these actions on materials have to be recognized in order to avoid faults and failures, defined as follows [5]:

- Fault (FR Panne, DE Fehlzustand): the condition of an item that occurs when one of its components or assemblies degrades or exhibits abnormal behaviour.
- Failure (FR Defaillance, DE Ausfall): the termination of the ability of an item to perform a required function. Failure is an event as distinguished from fault, which is a state.

The systematic examination of materials deterioration mechanisms which may cause faults and failures is called root cause failure analysis. An overview of the types of functional stress and environmental influences on materials, and the main deterioration mechanisms is given in Fig. 2.8 [6].

- Fracture—the separation of a solid body into pieces under the action of mechanical stress—is the ultimate failure mode. It destroys the integrity and the structural functionality of materials.
- Ageing results from all the irreversible physical and chemical processes that happen in a material during its service life. Thermodynamically, ageing is an inevitable process,

however its rate ranges widely as a result of the different kinetics of the single reaction steps involved.

- Electromigration is the transport of material constituents caused by the gradual movement of ions in a conductor due to the momentum transfer between conducting electrons and diffusing metal atoms. The effect is important in microelectronics and related structures.
- Corrosion is defined as an interaction between a metal and its environment that results in changes in the properties of the metal (ISO Standard 8044). In most cases the interaction between the metal and the environment is an electrochemical reaction.
- Biodeterioration denotes detrimental changes of materials and their properties and may be caused by microorganisms (bacteria, algae, higher fungi, basidiomycetes) as well as by insects (termites, coleoptera, lepidoptera) and other animals.
- Wear is the process of deterioration of a solid surface, generally involving progressive loss of substance due to relative motion between contacting bodies. Wear is a complex process that depends on various system factors and interrelationships.

Materials deterioration and failures of technical items are due to a complex set of interactions between functional loads and environmental influences that act on and within a system, and the materials that the system comprises. *Reliability* is defined as the probability that a technical item will perform its required function [4]. The failure probability of a material under load L originates from both the interaction of the statistical scatter of the applied load stress σ_L and the statistical scatter of material properties determining the macroscopic strength σ_S of the component. Both quantities are denoted by σ in accordance with mechanical stress and strength.

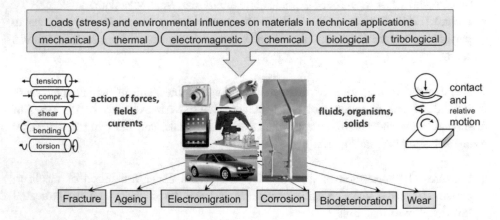

Fig. 2.8 Overview of materials deterioration mechanisms

- Stress means symbolically any physical process acting on the component. This can be a mechanical, a thermal, an electromagnetic, a tribological, or an environmental load. The load stress σ_L scatters due to varying conditions of loads and environmental influences.
- Strength means the ability of the component to withstand stress. Strength σ_S scatters due to scatters of the material properties and to scatters of manufacturing quality of the engineered component.

Failure takes place as soon as the stress distribution interferes with the strength distribution.

The results of various failure studies imply that two conceptual models of failure can be anticipated:

- *Overstress failures* are catastrophic sudden failures due to a single occurrence of a stress event that exceeds the intrinsic strength of a material. Overstress failures are, e.g., buckling of materials, electrical failures resulting in electrical discharge, or thermal overstress in a polymer beyond glass transition temperature.
- *Damage accumulation* is due to stress events that accumulate irreversibly and may lead to materials degradation and a gradual reduction of strength. Failure results when damage exceeds the endurance limit of the stressed material. Examples of damage accumulation are electronic electromigration, mechanical fatigue, tribological wear, and corrosion due to contaminants and weathering cycling.

Figure 2.9 illustrates three different phenomena of reliability and the occurrence of failure:

- (a) under normal conditions there is no interference of the stress and strain distributions and the failure probability is assumed to be zero,
- (b) under overstress the failure probability increases suddenly (catastrophic failure),
- (c) under damage accumulation the failure probability increases gradually (fatigue-type failure).

It follows that the reliable performance of materials, components and systems requires both, the avoidance of overstress through the appropriate choice of functional stress, and the control of damage accumulation by structural health monitoring.

Reliability Considerations

Reliability is defined as the probability that a technical item will perform its required functions without failure for a specified time period (lifetime) when used under specified conditions. As the lifetime of a device scatters, reliability is a quantity that can only be described by means of statistic tools. In order to describe the time dependence of a device's failure behavior, a reliability parameter called failure rate $k(t)$ is used. Simply speaking, it

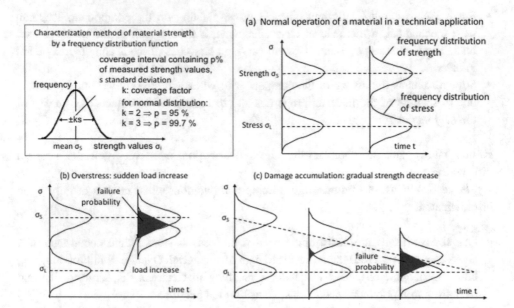

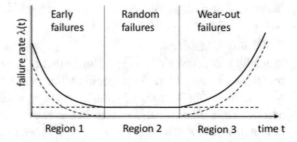

Fig. 2.9 Dynamic interference model of reliability. **a** Normal operation, **b** overstress failure, **c** damage accumulation

describes the (average) number of failures per time unit. In most cases, if k(t) is plotted against time, a typical bathtub-shaped curve is observed, Fig. 2.10.

The bathtub curve is categorized into three regions:

- Region 1 exhibits a falling failure rate and covers the so-called early failures (*infant mortality*). The origin for those failures in most cases is not related to material properties, but rather to the quality of the manufacturing process of the whole device.
- Region 2 is characterized by a constant failure rate and covers random failures that are not governed by a single failure mechanism. In this region, the device fails due to miscellaneous interactions with its environment, e.g. peak-like overloads, misuse, high temperatures and others.

Fig. 2.10 The characteristic failure rate as function of time (bathtub curve)

- Region 3 is the region where material-related failures start to dominate. Therefore, the somewhat imprecise phrase *wear-out failures* is commonly used for this phase. The failure modes in this region are often initiated by detrimental changes of the devices' material components, caused by the service loads applied to the device. The mechanisms leading to failures are called degradation mechanisms. Typical degradation mechanisms are mechanical fatigue, corrosion, wear, biogenic impact, and material-environment interactions.

2.7 Structural Health Monitoring and Performance Control

The objective of structural health monitoring (SHM) is to monitor and control the structural integrity of the components and the structure of technical systems. The following fundamental axioms for SHM have been formulated [7]:

I   All materials have inherent flaws or defects.
II The assessment of damage requires a comparison between two system states.
III Identifying the existence and location of damage can be done in an unsupervised learning mode, but identifying the type of damage present and the damage severity can generally only be done in a supervised learning mode.
IV Sensors alone cannot measure damage. Feature extraction through signal processing and statistical classification is necessary to convert sensor data into damage information.
V   The length- and time-scales associated with damage initiation and evolution dictate the required properties of the SHM sensing system.
VI There is a trade-off between the sensitivity to damage of an algorithm and its noise rejection capability.
VII The size of damage that can be detected from changes in system dynamics is inversely proportional to the frequency range of excitation.

The SHM process involves the observation of a system over time using periodically sampled dynamic response measurements from an array of sensors, the extraction of damage-sensitive features from these measurements, and the statistical analysis of these features to determine the current state of system health.

The application of technical diagnostics to the monitoring and diagnostics of structures is exemplified in Fig. 2.11.

Performance Control
The control of a functional parameter of a technical system with sensors and non-destructive evaluation (NDE) such that a significant change is indicative of a developing fault is called

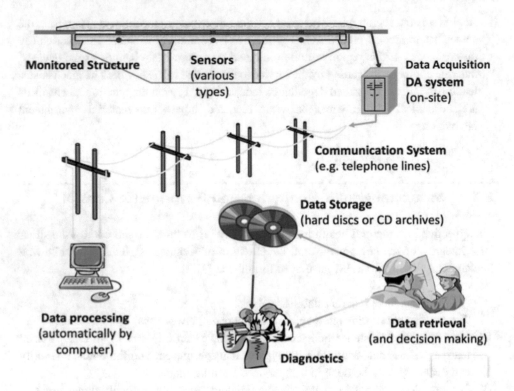

Fig. 2.11 Monitoring and diagnostics of structures with embedded sensors

Condition Monitoring [4]. The application of conditional monitoring allows maintenance to be scheduled, or other actions to be taken to avoid the consequences of failure, before the failure occurs. Examples of metrological condition monitoring parameters of technical systems of different machine types are exemplified in Fig. 2.12.

Combining the methodologies of structural health monitoring and performance control, a general scheme to support the functionality of technical systems results, which is shown in Fig. 2.13.

Parameter	Technical system (machine type)						
	Electric motor	Gas turbine	Pump	Compressor	Electric generator	Fan	Internal combustion engine
Temperature	★	★	★	★	★	★	★
Torque	★	★		★	★		★
Pressure		★	★	★		★	★
Speed	★	★	★	★	★	★	★
Vibration	★	★	★	★	★	★	★
Voltage	★				★		
Current	★				★		
Air flow		★		★		★	★
Fluid flow			★	★			
Fuel flow		★					★
Oil presueure		★	★	★		★	★
Input power	★		★	★	★	★	
Output power	★	★			★		★
★ indicates condition monitoring measurement parameter is applicable							

Fig. 2.12 Examples of condition monitoring parameters for different technical systems

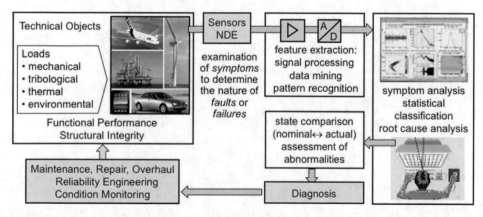

Fig. 2.13 A general scheme for the application of structural health monitoring and performance control to technical objects

References

1. von Bertalanffy, L.: General System Theory. Penguin, London (1971)
2. Wiener, N.: Cybernetics or Control and Communication. MIT Press, Cambridge, Mass (1948)
3. Feynman, R.P., Leighton, R.B., Sands, M.: The Feynman Lectures on Physics. Addison-Wesley, Reading (1963)
4. Czichos, H., Saito, T., Smith, L.: Springer Handbook of Metrology and Testing. Springer, Heidelberg, Berlin (2011)
5. BIPM: International Vocabulary of Metrology (VIM), 3rd edn. www.bipm.org (2008)
6. Czichos, H.: Measurement, Testing and Sensor Technology—Fundamentals and Application to Materials and Technical Systems. Springer, Heidelberg, Berlin (2018)
7. Worden, K., Farrar, C.R., Manson, G., Park, G.: The fundamental axioms of structural health monitoring. Philos. Trans. R. Soc. Math. Phys. Eng. Sci. **463**, 1639–1664 (2007)

Tribological Systems

3

The term *tribology* was coined 1966 in the UK after a comprehensive study on the enormous technological and economic importance of friction, lubrication and wear [1] with the following definition: *Tribology is the science and technology of interacting surfaces in relative motion and of related subjects and practices. The* collection of controlling factors and relationships affecting friction and wear is referred to as a *Tribological System* [2]: a structure of interacting surfaces in relative motion to perform a specified technical function.

Tribology has a "dual" character.

- On one hand, *interacting surfaces in relative motion* are necessary to realize technical functions like
    - the transmission of motion, forces and energy via interfacial traction,
    - the machining or forming of materials with workpiece-tool pairings,
    - transportation by wheel-rail or tire-road locomotion systems,
    - control of the flow of liquids and gases with pipelines and seals,
    - medical technologies supporting motion functions of the human body.
- On the other hand, *interacting surfaces in relative motion* may lead to friction-induced energy dissipation and wear-induced materials deterioration. The impact of friction and wear on energy consumption, economic expenditure, and CO_2 emissions has been outlined as follows [3]:
    - In total, about 23% of the world's total energy consumption originates from tribological contacts. Of that 20% is used to overcome friction and 3% is used to remanufacture worn parts and spare equipment due to wear and wear-related failures.

© The Author(s), under exclusive license to Springer Nature Switzerland AG 2022
H. Czichos, *Introduction to Systems Thinking and Interdisciplinary Engineering*,
Synthesis Lectures on Engineering, Science, and Technology,
https://doi.org/10.1007/978-3-031-18239-6_3

– By taking advantage of the new surface, materials, and lubrication technologies for friction reduction and wear protection in vehicles, machinery and other equipment, energy losses due to friction and wear could considerably be reduced.

– The largest energy savings through tribological expertise are envisioned in transportation (25%) and in the power generation (20%) while the potential savings in the manufacturing and residential sectors are estimated to be about 10%.

3.1   Characteristics of Tribological Systems

Basically, the science and technology of tribology is concerned with the optimization of technical systems from the study of friction and wear. This involves not only the properties of the materials in contact but also various system factors, such as the nature of the relative motion, the nature of the loading, the shape of the surface(s), the surface roughness, the ambient temperature, and the composition of the environment in which the wear occurs. As a consequence, some form of operational classification is needed to evaluate the material, mechanical, and environmental elements that can affect wear rates of a part.

Typical tribological systems (in short tribosystems) are exemplified in Fig. 3.1 [4].

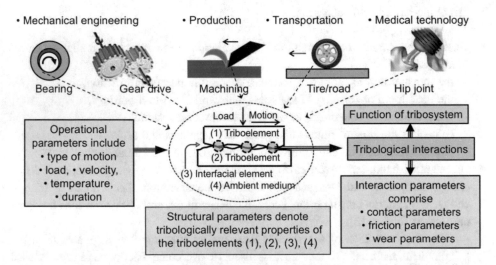

Fig. 3.1 Tribosystems in various technology areas and the basic groups of tribological parameters

Table 3.1 Overview of function of tribosystems and typical examples

Function of tribosystems	Examples
Motion transmission and control	Bearings, Joints, Clutches, Cam and followers, Brakes
Forces and energy transmission	Gears, Rack-and-pinion, Screws, Drives, Actuators, Motors
Data and images storage and print	Computer hard disc drives, Printing units
Transportation	Drive and control technology, Wheel/Rail, Tyre/Road
Fluid flow and control	Pipelines, Valves, Seals, Piston-cylinder assemblies
Mining	Dredging, Well drilling, Quarrying, Comminution
Forming	Casting, Drawing, Forging, Extrusion, Injection moulding
Machining	Cutting, Milling, Shaping, Boring, Grinding, Polishing
Medical technology	Medical Implants, Prosthetic devices, Dental technology

3.1.1 Function of Tribological Systems

There is a broad variety of technical functions which can only be realized with tribological systems. Examples are compiled in Table 3.1.

3.1.2 Structure of Tribological Systems

The structure of a tribological system consist of the components in contact and relative motion with each other [triboelements (1) and (2)], the interfacial element (3) between the contacting parts, and the ambient medium (4). Examples of the four basic triboelemnts are given in Fig. 3.2.

The structure of tribosystems may be either "closed" or "open". Closed means that all components are involved in a continuous, repeated, or periodical interaction in the friction and wear process, for example, as in bearings or a gear drive. Open means that the element in the tribosystem is not continuously involved in the friction and wear process and that a materials flow in and out of the system occurs, for example, workpieces in machining.

Basic structural parameters of triboelements (1) and (2) include:

- Geometric parameters: geometry, shape, dimenaions
- Chemical parameters: bulk chemical and molecular composition
- Physical parameters, such as thermal conductivity
- Mechanical parameters, such as elastic modulus, hardness, and fracture toughness.

Surface parameters of triboelements (1) and (2) are of special importance because tribological systems are based on *interacting surfaces in relative motion.* The characteristics

Tribosystem	Triboelements (1)	(2)	Interfacial element (3)	Ambient medium (4)	System structure
Bearing	Shaft	Bushing	Lubricant	Oil mist	Closed
Gear drive	Drive gear	Driven gear	Lubricant	Air	Closed
Machining	Tool	Workpiece	Cutting fluid	Air	Open
Tire/Road	Tire	Road	Moisture	Air	Open
Hip joint	Femur	Capsule	Synovial	Tissue	Closed

Bearing   Gear drive   Machining   Tire/road   Hip joint

Fig. 3.2 Examples of the structural elements of tribosystems

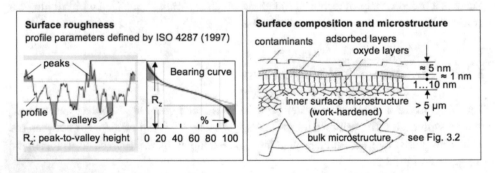

Fig. 3.3 Surface characteristics: roughness, subsurface composition, and microstructure

of the topography and the composition and microstructure of surfaces are depicted in Fig. 3.3.

3.1.3 Operational Parameters of Tribological Systems

The basic operational parameters of tribological systems are:

- **Type of motion**, classified in terms of sliding, rolling, spin, and impact and their possible superposition. The kinematics can be continuous, intermittent, reverse, or oscillating.

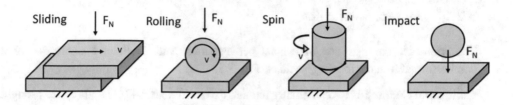

- **Normal load** (F_N), defined as the total force (including weight) that acts perpendicular to the contact area (A) between the contacting elements, where the contact pressure, p, is given by $p = F_N / A$.
- **Velocity** (v), specified as the vector components and the absolute values of the individual motions of the contacting elements.
- **Temperature** (T) of the structural components at a stated location and time. In addition to the operating (steady-state) temperature, the friction-induced temperature rise (average temperature rise and flash temperatures), must be measured or estimated on the basis of friction heating calculations.
- **Duration** (time t) of the operation, performance or test.
- The basic set of operational parameters can be expressed as: {$\mathbf{F_N}$, $\mathbf{v}$. $\mathbf{T}$, $\mathbf{t}$}.

3.2   Interactions in Tribological Systems

Interaction parameters characterize the action of the operational parameters on the structural components of a tribological system. An overview is given in Fig. 3.4 [2].

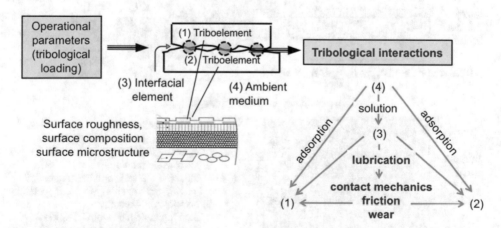

Fig. 3.4 Overview of interaction parameters of tribological systems

3.2.1 Contact Mechanics

The geometric configuration of contact between two parts is either a conformal or a counterformal (nonconforming) contact, Fig. 3.5.

Conformal Contact: If two nominally flat and dry solid materials (1) and (2) are brought into static contact under the action of a normal load, the touching asperities of this tribocontact deform elastically or plastically under the given load. The dominating contact deformation mode is governed by a deformation criterion (the so-called plasticity index), which depends on the deformation properties and the parameters of the surface topography of the contacting bodies (1) and (2) [5]. The summation of individual contact spots gives the "real area of contact", A_r, which usually is much smaller than the "apparent geometrical area" A_0 of the contact. The real area of contact depends in the elastic case primarily on the ratio of F_N to the elastic modulus of both bodies, and in the plastic case on the ratio of F_N to the yield pressure or the hardness of the softer of the contacting bodies.

If, in addition to the normal load, a tangential force is introduced, a junction growth of asperity contacts may occur. The microscopic interaction forces between contacting solids include, at least in principle, all those types of atomic and molecular interaction that contribute to the cohesion of solids, such as metallic, covalent, and ionic, that is, primary chemical bonds (short-range forces), as well as secondary van der

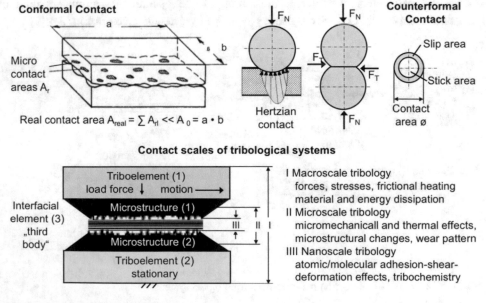

Fig. 3.5 Characteristics of tribological contacts

Waals bonds (long-range forces) [5]. These surface forces depend in a complicated manner on the physico-chemical nature of the structure and composition of the outermost surface layers and contaminants of the material of triboelements The interfacial local stress-adhesion-deformation-shear processes in the stochastically distributed micro-contacts cause—independent from the nominal geometric contact area—the frictional resistance against motion.

Counterformal Contact. For curved bodies, the macroscopic elastostatic contact situation is described by the well-known Hertzian equations. If a tangential force $F_T < F_N \cdot f_0$ (where f_0 is the static coefficient of friction) is additionally applied to a static Hertzian contact, a separation of the Hertzian contact area in a slip area and stick area takes place before a macroscopic horizontal relative motion between body (1) and body (2) occurs. In the case of macroscopic relative motion, the superposition of normal forces, F_N, and frictional forces $F_F > F_N \cdot f_0$ (perpendicular to F_N) in a tribocontact leads to complex stress distributions in the contacting bodies. The multiaxis stress condition may be converted into a uniaxial tensile stress in such a way that the uniaxial tensile load affects the material to the same extent as the multiaxial stress state.

The contact scales of tribological systems are illustrated in the lower part of Fig. 3.5.

Macroscale tribology refers to tribosystems with dimensions from millimeters up to meters; sliding speeds are in the range of approximately 1 mm/s to more than 10 m/s. The function of macroscale engineering components—to transmit motion, forces, and mechanical energy—is governed by dynamic mass/spring/damping properties of the interacting elements.

Microscale tribology typically involves devices that are 100–1000 times smaller than their macroscale analogs. Thus, the volume of components on this scale is reduced by a factor of at least 10^6. As an example, the technology of microelectromechanical (MEM) systems is an interdisciplinary technology dealing with the design and manufacture of miniaturized machines, with the major dimensions at the scale of tens to hundreds of micrometers. Masses and inertias of MEM components rapidly become small as size decreases, whereas surface and tribological effects, which often depend on area, become increasingly important.

Nanoscale tribology involves phenomena from the submicroscopic to the atomic scale. It is not possible to link the test results obtained at the nanoscale with friction and wear phenomena at the macroscopic scale, mainly because atomistic models are not directly scalable [6]. For the diagnosis of friction and wear test data it must be noted that the conditions in which nanotribology and macrotribology tests take place—using, for example atomic force microscopes (AFMs) for nanotribology and pin-on-disc testers for macrotribology—are very different. AFM tips induce stresses in the gigapascal range, whereas macroscale testers operate in the kilopascal range. Atomic force microscopes typically allow sliding amplitudes of only a few micrometers, whereas macrotesters sliding amplitudes range from hundreds of micrometers up to meters.

3.2.2 Fundamentals of Friction

Friction is the resistance to motion, when under the action of an external force one body moves or tends to move relative to another body. The quantity to be measured to characterize friction – in the terminology of metrology called the *measurand* of friction—is the friction force F_F. The friction force is a vector to be characterized by its direction and its quantity value. The coefficient of friction f is a dimensionless number, defined as the ratio between the friction force F_F and the normal force F_N acting to press the two bodies together $f = F_F/F_N$. The average frictional power P_F is defined as frictional energy E_F divided by the operating duration, t (that is, $P_F = E_F/t$). For sliding friction, $P_F = F_F \cdot v = f \cdot F_N \cdot v$.

The physics of friction has been explained by Richard Feynman in his famous *Feynman Lectures on Physics* [7], see information box.

Physics of Solid Friction*

"Dry sliding friction occurs when one solid body slides on another. A force is needen to maintain motion. This force is called a frictional force and its origin is a very complicated matter. Both surfaces of contact are irregular, on the atomic level. There are many points of contact where the atoms seem to cling together, and then, as the sliding body is pulled along, the atoms snap apart and vibration ensues. As the slider snaps over the bumps, the bumps deform and then generate waves and atomic motions, and, after a while, friction-induced heat in the two bodies".

Sliding Friction Model on the Atomic Scale

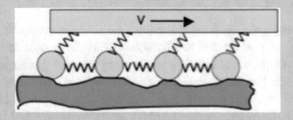

Quantity values of the friction coefficient. "Tables with data of friction coefficients for materials, like "steel on steel" or "copper on copper", are all false. The friction is not due to "copper on copper" because the surfaces in contact are not pure copper but are mixtures of oxides and other impurities. It is impossible to get the right coefficient of friction for pure metals because if ultraclean pure metal surfaces are brought into contact the interfacial atomic forces became cohesive and the two pieces stick together. The friction coefficient which is ordinarily less than unity for reasonably hard surfaces becomes several times unity".

This concise statement was confirmed by friction measurements made by NASA**. Depending on the interfacial and ambient conditions, the following data were measured for the friction coefficient f of „copper on copper":

f = 0.08, measured under boundary lubrication (mineral oil),
f = 1.0, measured as solid friction in air,
f > 100, measured as solid friction in vacuum (10^{-10} Torr).

*The Feynman Lectures on Physics, Addison, Wesley, 1963, Chapter 12-2 Friction.

**Buckley [5].

Friction Mechanisms
An overview of the mechanism of friction can be obtained from considering the energy balance (energy dissipation event) of solid friction. The mechanical energy associated with solid friction between two solid triboelements (1), (2) involves the following, Fig. 3.6:

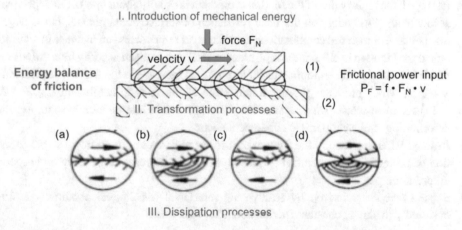

Fig. 3.6 Overview of the basic mechanisms of friction

I.   Introduction of mechanical energy: formation of the real area of contact, junction growth
     at the onset of relative motion
II. Transformation processes:
     (a) adhesion and shear,
     (b) plastic deformation,
     (c) abrasion,
     (d) hysteresis and damping
III. Dissipation processes:
     (i)   thermal processes,
     (ii) absorption in (1) and (2) with residual stresses, generation of point defects and
           dislocations, stacking fault energy
     (iii) emission, e.g. heat, debris, noise, triboluminescence, exo-electrons.

The occurrence of friction mechanisms depends on the structural and operational parameters
of the tribosystem in question. All of the partial processes of friction in Fig. 3.6 have been
experimentally observed in various studies [4].

Experimental Example: Dry Sliding Friction of Steel on Steel
The simplest type of tribotest is to subject a given tribosystem to a defined constant set of the
basic operational parameters {load F_N, velocity v, temperature T} and to measure friction
or wear quantities as functions of time t only.

A typical graph for the course of the friction coefficient $f = F_F/F_N$ of a dry metal/metal
sliding system as function of time t is shown in Fig. 3.7. The friction-time curve consists in
a simplified presentation of four stages [4].

* At stage I, the initial value of the friction coefficient is usually about $f_0 \approx 0.1$. It dependents
  at low loads, F_N, mainly on the shear resistance of surface contaminants, but is largely
  independent of material combinations. Surface layer removal and an increase in adhesion
  due to the increase in clean interfacial areas as well as increased asperity interactions and
  possible wear particle entrapment lead to a gradual increase in the friction coefficient.
* Stage II, which produces the maximum value of the friction coefficient ($f_{max} \approx 0.3$
  ...1.0 for most metal pairs), is reached when maximum interfacial adhesion, asperity
  deformation, and wear particle entrapment occur.
* In stage III, a decrease in the friction coefficient may occur due to the possible formation of protective tribochemical surface layers and a decrease in plowing and asperity
  deformation processes.
* Stage IV is characterized by steady-state interfacial tribological conditions leading
  eventually to almost constant friction coefficient values.

It should be noted that Fig. 3.7 shows a simplified smoothed friction graph, which in practice
may be overlapped by short-term fluctuations, friction peaks, or stick–slip effects. In any

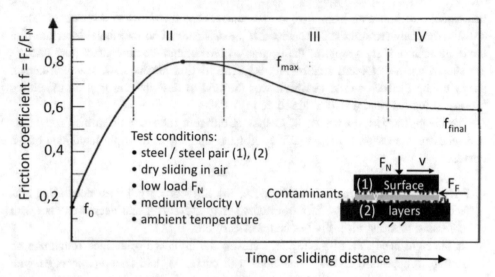

Fig. 3.7 A Friction-time graph with the initial friction coefficient (f_0), the maximum friction coefficient (f_{ma}), and the friction coefficient at the end of test (f_{final})

case, the friction-time behavior should be characterized by three characteristic values: (a) the initial friction coefficient, f_0, (b) the maximum friction coefficient, f_{ma}, (c) the friction coefficient at the end of test, f_{final}.

3.2.3 Friction and Lubrication

Friction forces under lubricated conditions are mostly lower than those of solid friction and depend on the shear resistance of interfacial boundary films and on the rheology of the lubricant. The basic tribological parameter of friction and lubrication is the dimensionless **film-thickness to roughness ratio λ** [4].

The experimental verification of the dependence of friction on the film-thickness to roughness ratio λ is demonstrated in Fig. 3.8 [7]. A cylinder-on-flat model configuration with a constant load F_N was used as test system and the tests were performed with three decreasing lubricant viscosity values at oil bath temperatures of 20 °C (A), 30 °C (B), and 40 °C (C). The sliding velocity v was stepwise increased, kept constant at every data point for 5 s, and the friction force F_F was measured with a force sensor. Simultaneously, an information on the nature of the contact (metal/metal or metal/lubricant/metal) was obtained by measuring the electrical contact resistance $R_{eletr.}$, between the sliding triboelements. A contact resistance of $R_{electr.} < 10\ \Omega$ indicates metal/metal contact, and $R_{electr.} > 50\ k\Omega$ indicates separation of the sliding surfaces by the lubricant film. With a fast signal discrimination technique, the no-contact time fraction was determined. Thus, a distinction between solid/solid contact and the existence of an interfacial lubricant film could be achieved. The graphs of the friction coefficient and the no-contact time fraction are shown on the left-hand side of Fig. 3.8. The friction graphs show the well-known shape of the Stribeck curve. (Stribeck was the first to describe the friction of sliding lubricated journal bearings with this curve [8].)

The single Stribeck curves A, B, C show a different influence of surface roughness and lubricant viscosity on friction at the left hand side and at the right hand side of the curves:

- Friction decreases at the left hand side of the Stribeck curves with decreasing surface roughness ("run-in" effect from curve A to curve A*) and decreases also with increasing lubricant viscosity (sequence of curves C, B, A).
- At the right hand side of the Stribeck curves, the influence of surface roughness on friction disappears (curve A* coincides with curve A), and friction increases with increasing lubricant viscosity (sequence of curves C, B, A).

A generalized Stribeck curve could be obtained by combining the experimentally determined friction and no-contact data with calculated film-thickness data based on the elastohydrodynamic (EHD) lubrication theory and on the values of the operational and materials parameters of the tests. The generalized Stribeck curve characterizes the

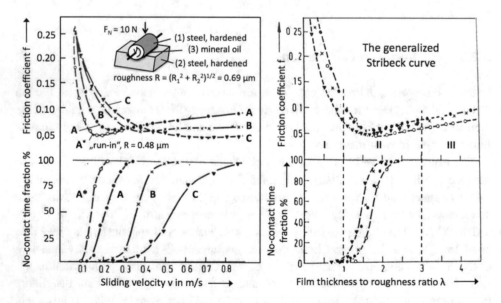

Fig. 3.8 Experimentally determined Stribeck curves and the generalized Stribeck curve

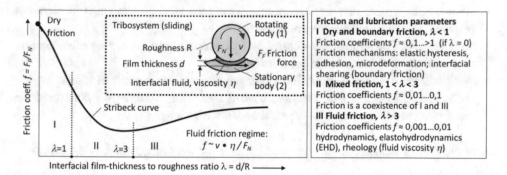

Fig. 3.9 Overview of the regimes of friction and lubrication

main friction and lubrication regimes as function of the dimensionless film-thickness to roughness ratio λ, see Fig. 3.9.

3.2.4 Fundamentals of Wear

Wear is the process of deterioration of a solid surface, generally involving progressive loss of substance due to relative motion between contacting bodies, that is, the interacting elements of a tribosystem. The measures and units used to quantify wear are illustrated in Fig. 3.10:

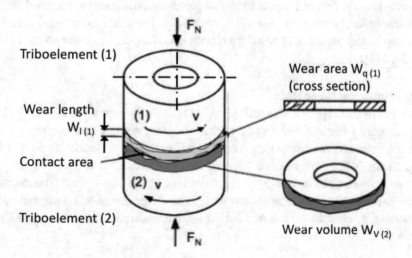

Fig. 3.10 Wear measuring quantities

- Wear length, W_l: one-dimensional changes in the geometry of interacting triboelements perpendicular to their common contact area.
- Wear area, W_q: two-dimensional changes of cross sections of interacting triboelements perpendicular to their common contact area.
- Wear volume, W_V: three-dimensional changes of geometric regions of interacting triboelements adjacent to their common contact area. Wear volumes are connected via density or specific gravity with wear masses or wear weights.

A further wear parameter is the wear coefficient, k (or wear rate), which is defined as

$$k = W_V/F_{Ns}\ (mm^3/N\,m)$$

The wear coefficient k gives a relation between the wear volume and the operational parameters load and sliding distance causing wear. The denominator (in units of N m) of the wear rate (k) multiplied by the coefficient of friction F_F/F_N gives the value of dissipated energy (in units of MJ).

Wear Mechanisms

The chain of events leading to wear is illustrated in Fig. 3.11. Tribological loading comprises the actions of contact mechanics, relative motion, and contacting matter, activated by frictional energy. The action of contact forces and stresses in combination with relative motion trigger the wear mechanisms of surface fatigue, abrasion, and the materials degradation processes listed in the left part of Fig. 3.11. The contacting matter initiates—together with interfacial and ambient media—the wear mechanisms of tribochemical reactions and adhesion in combination with debris formation processes. All these processes and their interference lead to wear: surface damage and wear particles. The combined materials degradation effect of wear and corrosion is called tribocorrosion. Wear surface appearances are shown in Fig. 3.12 [4].

Wear Regimes and Wear Data

An overview of the range of wear data is given in Fig. 3.13. Depending on the interfacial state of a tribosystem—lubricated sliding, unlubricated sliding, wear by hard particles—the wear coefficient k can vary over more than six orders of magnitude. A wear coefficient of 10^{-6} is conventionally indicating the border between tolerable removal of surface boundary layers to mild and severe wear. Abrasive hard particles cause even higher wear. The lowest wear can only be achieved when the triboelements (1) and (2) are completely separated by a fluid film, realized by elastohydrodynamic (EHD) lubrication, or hydrostatic or hydrodynamic lubrication.

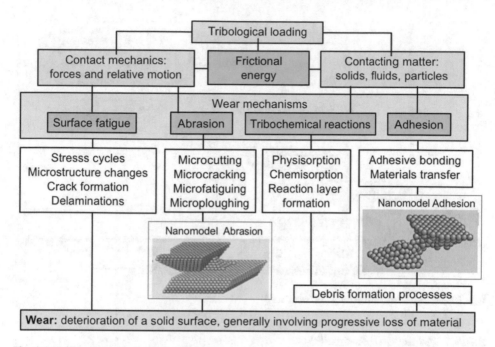

Fig. 3.11 The chain of events leading to wear

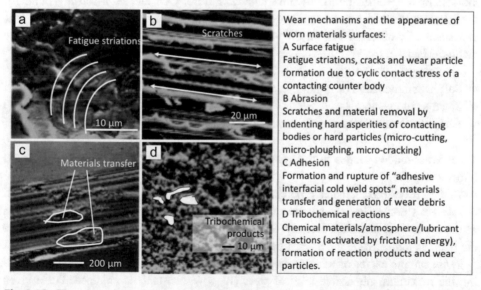

Fig. 3.12 The appearance of worn surfaces (scanning electron microscopy, SEM)1

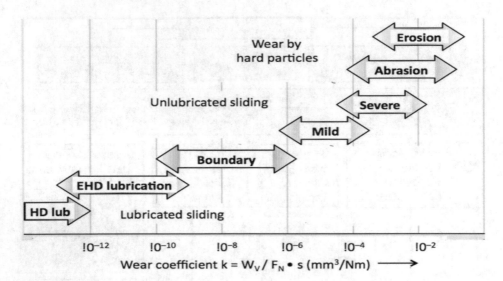

Fig. 3.13 Overview of the range of wear data

3.2.4.1 Wear and Reliability

Reliability describes the ability of a system or component to function under stated conditions for a specified period of time without failure. Consider as a starting point for the reliability considerations of tribosystems the wear behavior as function of time. It is experimentally observed that for time-dependent wear, three different wear stages can be distinguished:

- Self-accommodation or running-in wear
- Steady-state wear
- Self-acceleration (catastrophic damage) of wear.

These wear-mode changes in the system behavior may follow each other with time, as indicated in Fig. 3.14.

In Fig. 3.14, W_{lim} denotes a maximum admissible level of wear losses. At this level the system structure has changed in such a way that the functional input–output relations of the system are disturbed severely. Repeated measurements show random variations in the data, as indicated by the dashed lines.

From sample curves of wear, a probability density function, $f(t)$, of the time for reaching the maximum admissible level of wear ($W_{\text{lim}} = $ constant) is obtained. For a given time, t_0, the shaded area under the curve $f(t)$, i.e. the value of $F(t_0)$, is a measure of the probability that the system fails within the time $t < t_0$.

For the modeling of the distribution of measured wear data and the estimation of reliability data, statistical distributions may be used. If for a given tribosystem the failure mode

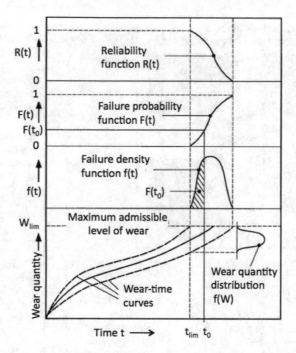

Fig. 3.14 The dependence of wear on time, and failure and reliability functions

and the failure distribution can be determined, this knowledge can be used to improve the reliability of the system. For tribosystems failing as a consequence of wear processes, the failure behavior may be characterized by the Normal distribution, the Weibull distribution, or the Gamma distribution. Their mathematical functions and statistical parameters are well documented in the statistical literature.

3.3 Condition Monitoring of Tribological Systems

The most widely used methods for Condition Monitoring of Tribological Systems are Vibration Monitoring and Non-destructive Evaluation.

Vibration is defined as a periodic motion about an equilibrium position. Its duration and magnitude depend upon the degree of damping the effected materials possess and the phase relationships between the mechanism that perturbs the system and the response that is obtained. Vibration may be forced through unbalance, rub, looseness and misalignment, or freely self-excited through internal friction, cracking and resonance.

Once generated, vibration can be transmitted from its source to other components or systems. When it reaches unacceptable levels, tribological wear and tear processes are accelerated, which in turn initiate various failure mechanisms. Hence, by monitoring for

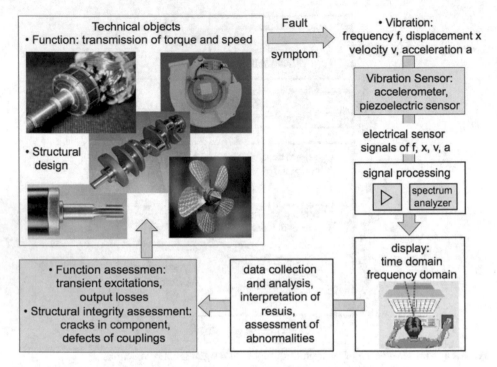

Fig. 3.15 Methodology of vibration analysis exemplified for rotary machine equipment

the presence and change of vibration patterns through the methods of signature analysis or shock pulse, the consequences of avoidable breakdowns can be prevented. Figure 3.15 illustrates the general methodology of vibration monitoring.

An application example of condition monitoring of a bearing is given in Fig. 3.16. It illustrates the changes that a frequency spectrum—obtained from vibration accelerometer readings—will undergo as an outer race spall develops for a bearing with a specific outer race defect frequency [9].

Nondestructive Evaluation

A wide variety of inspection and detection techniques are available to provide users with the speed, accuracy, and cost-efficiency needed to probe, identify, and diagnose features of import in validating quality control and product fitness. Table 3.2 gives an overview of established NDE methods for technical diagnostics.

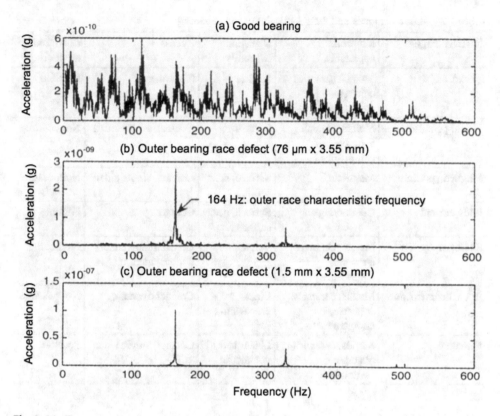

Fig. 3.16 Frequency spectra of failing bearing

3.4 Testing of Tribological Systems

Tribotesting has the task to investigate the influence of friction and wear on the function and structure of tribological systems. Basic objectives are:

- Evaluation of the impact of friction and wear on the performance, reliability, and safety of engineering tribosystems or tribocomponents.
- Quality control of tribocomponents.
- Characterization of the trihological behavior of materials and lubricants.
- Investigation of basic tribological processes, and friction-induced energy losses or wear-induced materials losses.

Laboratory friction and wear experiments must be based—as all experiments in science and technology—on the principles of measurement and testing.

Table 3.2 NDE techniques and their applicability and capability

NDE technique	Materials applicable	Detection capability	Access	Condition monitoring
Visual	Metals, ceramics, concrete, composites	Surface	Noncontact	Possible
Liquid penetrant	Metals, ceramics, concrete, composites	Surface	Contact, single sided	Not possible
Magnetic particle	Magnetic	Surface, near surface	Contact, single sided	Not possible
Eddy current	Conducting	Surface, near surface	Contact	Possible
Radiography	Metals, ceramics, concrete, composites	Volumetric	Contact/noncontact double sided	Possible
Acoustic emission	Metals, ceramics, concrete, composites	Linear propagating	Contact dynamic	Not possible
Ultrsonic	Metals, ceramics, concrete, composites	Linear and volumetric	Contact, single sided	Possible

Categories of Tribometry

Depending on the structure and function of the tribological system under consideration, the tests can be grouped into six basic categories, see Fig. 3.17 [10].

Tests of category I are performed with actual tribomachinery under practical conditions.

Clearly, these tests represent the real performance of the object under study. However, full-scale field tests are usually very expensive. The broad spectrum of the practical operating conditions often cannot be adequately characterized, and considerable efforts are necessary to confirm test results in a statistical manner in repeated tests.

In tests of category II–IV, in which actual tribomachinery, tribosystems, or tribocomponents are studied in well-defined bench tests, test results can be related to the actual triboengineering structures. An important consideration of these tests is the appropriate choice of operational conditions, which may be simplified, simulated, or accelerated as compared with the often unknown broad spectrum of the practical full-scale operational conditions. Sometimes attempts are made to determine the actual operational conditions of the full-scale filed performance with sensors and to simulate these conditions in less expensive bench tests. Because tribotests of categories I–IV differ in scope and testing

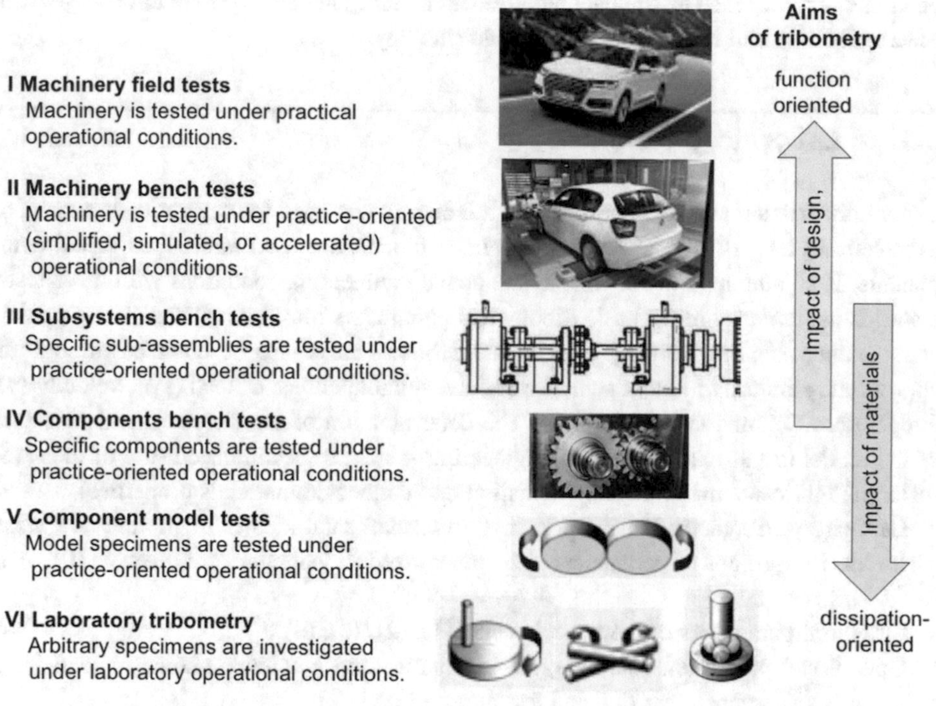

Fig. 3.17 Categories of tribotesting

conditions, the techniques and procedures to be applied for the determination of friction and wear quantities must be related to the individual tests and cannot be generalized.

In tests of category V, model test specimens are used rather than actual tribocomponents. If an attempt is made to simulate actual triboengineering conditions with these tests, a sufficient similarity in the basic tribological parameters must be realized. These parameters, include:

- The determination of the physico-chemical nature of the model test specimens and the triboengineering system to be simulated (use of the same materials/lubricant/environment combination and connected materials properties)
- Similarity of the contact and lubrication mode—characterized by similar counterformal or conformal contact conditions, and a similar working point in the Stribeck curve—to be obtained through an appropriate similarity in the kinematics, and the suitable choice of load, F_N, velocity, v, temperature, T, and test duration.

Tests of category VI are used primarily in fundamental studies of friction and wear processes. The conditions of these tests are often selected to study specific tribological phenomena rather than to simulate real triboengineering behavior. These tests range from Macrotribology and Microtribology to Nanotribology.

3.5   Laboratory Tribometry

Laboratory friction and wear experiments are mainly designed for categories V and VI of tribotests. In tests of category V, test-specimen models are used rather than actual components. If an attempt is made to simulate actual engineering conditions with these tests, a sufficient similarity in the basic tribological parameters must be realized. These parameters include: similar working point in the Stribeck curve—to be obtained through an appropriate similarity in the kinematics and the suitable choice of load (F_N), velocity (v), temperature (T), and test duration (t). The determination of the physicochemical nature of the model test specimens and the engineering system to be simulated (use of the same materials/lubricant/environment combination and connected materials properties).

Laboratory tribometry is used primarily in fundamental studies of friction and wear. There are four groups of parameters to be recognized in laboratory tribometry [10]

- Structural parameters of the triboelements (1), (2), (3), (4).
- Operational parameters including the kinematics and a suitable choice of load (F_N), velocity (v), temperature (T), and test duration (t).
- Quantities to be measured (denoted as "tribomeasurands" in the terminology of metrology).
- Surface characteristics of the test specimen (1) and (2) before and after testing.

A general scheme for laboratory measurement of friction and wear is shown in Fig. 3.18.

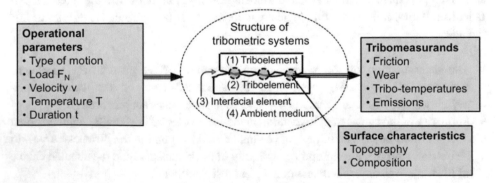

Fig. 3.18 General scheme for laboratory measurement of friction and wear

3.5.1 Methodology of Tribometry

The methodology of laboratory tribotests to study friction and wear with measurement, testing and sensor technology consists basically of the following steps [10].

1. Choose a suitable test configuration for the test specimens. Specify the geometry of the test configuration, the bulk materials properties and metallurgical description, and the surface topographies and characteristics. Note that the interfacial tribological processes act differently on each of the two triboelements depending on the contact-area to wear track ratio, see Fig. 3.19.
2. Characterize the interfacial element (for example, the lubricant) and the environmental medium or atmosphere in terms of their chemical nature, composition, and chemical and physical properties.
3. Choose a suitable set of the operational parameters, including type of motion, load F_N, velocity v, temperature T, and test duration t.
4. Perform the tests as functions of varied structural parameter of the triboelements (for example, hardness or roughness) and/or operational parameter (for example, load cycles or velocity variations). The conditions of the tests, such as the time dependence of operational parameters, should be controlled by appropriate detectors or sensors and supported by on-line computer techniques.
5. Measure the basic tribomeasurands, namely friction force and wear quantities. If of interest determine triboinduced thermal quantities (for example, friction-induced temperature rise), or triboinduced physical emissions (e. g. noise, vibrations), or chemical emissions (e.g. CO_2) or wear particle debris.

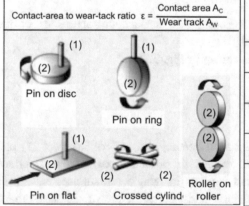

Contact-area to wear-tack ratio $\varepsilon = \dfrac{\text{Contact area } A_C}{\text{Wear track } A_W}$		Characteristics of triboelements	
		Element (1) stationary, $\varepsilon = 1$	Element (2) moving, $\varepsilon < 1$
		Permanent tribocontact	Intermittent tribocontact
		Quasi-static contact stress	Cyclic contact stress
		Permanent friction heating	Cyclic friction heating when A_W passes A_C
		Permanent action of wear mechanisms	Intermittent action of wear mechanisms
		No chemical influence of ambient medium in A_C	Chemical influence of the ambient medium on the area $A_W - A_C$

Pin on disc · Pin on ring · Pin on flat · Crossed cylind · Roller on roller

Fig. 3.19 The contact-area-to-wear-track ratio indicates the different action of tribological processes on the tribocomponents

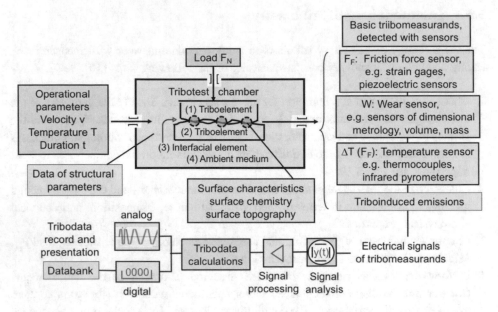

Fig. 3.20 General scheme for laboratory measurement of friction and wear

6. Investigate the surface characteristics of worn surfaces with respect to surface chemistry and surface topography (and wear particles) in order to identify wear mechanisms.
7. Record the tribodata and present the results of tribotesting in an appropriate manner. The full set of data from the tribosystems and the resulting tribological quantities shall be saved in a tribological data archive system.

In addition to the determination of tribomeasurands, surface characteristics of the interacting materials have to be determined.

The scheme of a tribometer prototype is shown in Fig. 3.20.

3.5.2 Surface Investigations

After a tribotest, the worn surfaces (and subsurface regions) of triboelement (1) and triboelement (2) should be analyzed with respect to surface topography and surface composition. Diagnosis of wear mechanism requires imaging worn surfaces as well as information on the local chemical composition and microstructure of the near-surface material. In addition, subsurface defects such as cracks or pores should be identified. Various techniques are employed to study tribological surfaces and subsurface regions. These include imaging techniques and methods of microstructural and compositional analysis.

The methods vary in the information they provide in both lateral extent and depth into the surface. In addition, analysis of wear debris and wear particles is performed.

Surface Chemical Analysis. Besides the bulk materials chemistry, surface characteristics of materials are of great significance. The main techniques of surface chemical analysis are [10]:

- EDX: energy-dispersive x-ray spectroscopy
- ESCA, XPS: electron spectroscopy for chemical analysis, x-ray photoelectron spectroscopy
- AES: auger electron spectroscopy
- SIMS: secondary ion spectrometry.

An overview of the instrumental features of this techniques are given in Table 3.3.

AES is excellent for elemental analysis (exception H, He) at spatial resolutions down to 10 nm, and XPS can define chemical states down to 10 μm. Both analyze the outermost atom layers and, with sputter depth profiling, layers up to 1 μm thick. Dynamic SIMS incorporates depth profiling and can detect atomic compositions significantly below 1 ppm. Static SIMS retains this high sensitivity for the surface atomic or molecular layer but provides chemistry related details not available with AES or XPS.

Table 3.3 Surface chemical analysis techniques

Analytical task	Method	Resolution	Comment
Depth profiling for elements and compounds by ion sputtering	ESCA	10 μm lateral, 1 nm in-depth	Principally, ion sputtering is destructive and information is reduced in elemental analysis
	AES	10 nm lateral, 1 nm in-depth	
	SIMS	50 nm lateral, 1 nm in-depth	
Layer thickness measurement	ESCA	1 cm^2 lateral	Thickness range: 1 monolayer to 10 nm oxydlayer thickness. Sample must be flat
	EDX	0.5–3 μm lateral	Thickness range: 10 nm to 1 μm. Layer density must be known
Chemical state analysis	ESCA	10 μm lateral	Analysis relies on databases (chemical shifts in electron spectroscopy, characteristic fragment patterns for SIMS)
	AES	10 nm lateral	
	SIMS	100 nm lateral (metals and semiconductors), 1 μm (organics)	

Surface Topography Analysis. The basic methods for surface topography measurements arestylus profilometry, atomic force microscopy (AFM), optical interferometry, and scanning electron microscopy (SEM) (Ref. [4]). Figure 3.21 gives an overview on data and experimental principles. The basic features of the techniques for surface topography analysis can be summarized as follows [10]:

- Stylus profilometry: The pick-up draws a stylus (diamond tip, 60° or 90°, radius of 1–10 μm) over the surface, and an analog electrical signal of the surface profile is produced by a piezoelectric, inductive or laser interferometric transducer. The spatial resolution achieved by this method, generally in the range of 2–20 μm, is limited by tip geometry and local plastic deformation.
- Atomic force microscopy: In an AFM, a sharp tip with a radius of approximately 5–20 nm, mounted on a very soft cantilever with a spring constant in the range of 1 N/m, is scanned over the surface by a xyz actuator with a resolution of much less than a nanometer and a dynamic range on the order of 10 μm in the z direction and up to 100 μm in the x- and y-directions.
- Optical interferometry: Interference microscopy combines an optical microscope and an interferometer objective into a single instrument. These optical methods allow fast noncontacting measurements on essentially flat surfaces. Interferometric methods

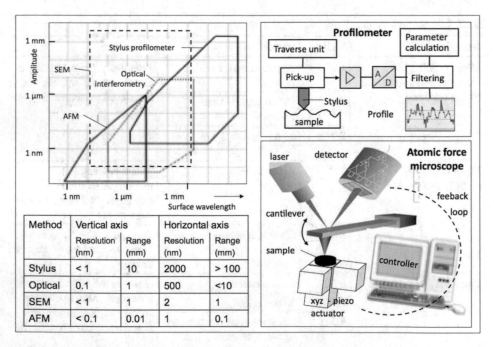

Fig. 3.21 Techniques of surface topography analysis

offer a subnanometer vertical resolution, being employed for surfaces with average roughnesses down to 0.1.

- Scanning electron microscopy (SEM) can provide both topographic (for example, from secondary electron detection) and compositional (such as the mean atomic number from back-scattered electrons) information. SEM is an important tool to identify wear mechanisms from the appearance of worn material surfaces as illustrated above in n Fig. 3.21.

3.6   Presentation of Tribodata

The presentation of the results of tribometry can be done in various ways to describe the system-dependent of tribodata:

- The time-dependence of friction is illustrated in Fig. 3.7.
- The time-dependence of wear is illustrated in Fig. 3.14.

3.6.1   Dependence of Tribodata on Operational Parameters

In addition to graphs illustrating the time-dependence of friction and wear, tribodata may be presented as function of the operational parameters {load force F_N, velocity v, temperature T}. The dynamic action of operational parameters may change the initial nominal structural parameters of materials (e.g. surface roughness, composition, microstructure). These changes influence in turn the friction and wear behavior of materials and their tribodata expressed in tribographs.

The dependence of the friction coefficient on the operational parameters {F_N, v,} for two different temperature T is illustrated in Fig. 3.22 for the example of pin-on-disc test system consisting of a polytetrafluorethylene, (PTFE) pin sliding on a smooth steel surface [10] Two essentially different friction regimes are observed:

(a) a low friction regime with a friction coefficient of f ≈ 0.03 at low sliding velocities v < 1 cm/min, and

(b) a regime of increasing friction values, f ≈ 0.03 → 0.3 for v > 1 cm/min.

The friction graphs of Fig. 3.22 show that at high contact pressures and low sliding velocities, low friction results. Optical micrographs of the steel surfaces indicate a thin polymer film adherent at the steel surface. Friction is low because the polymer chains orient themselves in the sliding direction and can be easily sheared. In the high friction

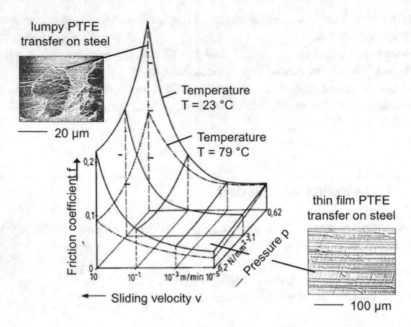

Fig. 3.22 Example of the dependence of the friction coefficient on the set of operational parameters $\{F_N, v, T\}$

regime, the polymer chains have not sufficient time to orient themselves in the sliding direction. Higher frictional shear forces are needed and the polymer transfer to the steel surface consists of polymer lumps.

3.6.2   Dependence of Tribodata on Structural Parameters

Tribodata are not correlated in a simple way with nominal structural properties of materials (like modulus or strength) because in its initial stages, friction and wear depend on contaminants and surface layers. If contaminants and outer surface layers are rubbed away, for example by an abrasive material, a correlation of the wear resistance with the hardness of the tribologically stressed material can be observed. Figure 3.23 shows simplified tribographs of the dependence of the abrasive resistance on the hardness for various classes of materials [10]. It must be emphasized that the trends indicated are found only when the hardness of the abrasive is at least about 1.3 times that of the material.

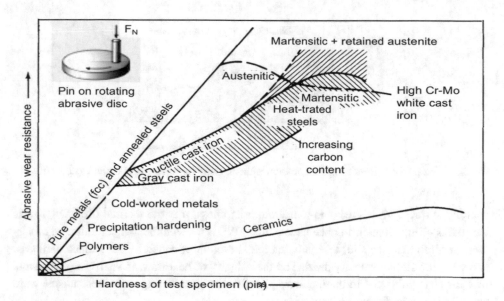

Fig. 3.23 Abrasive wear resistance of different materials as function of bulk hardness

3.6.3 Influence of Ambient Media on Tribodata

Ambient media can significantly affect the tribological behavior of materials. Figure 3.24 shows an example of the pronounced influence of humidity on the tribological behavior for two material pairings: (a) steel/silicon carbide (SiC) and (b) steel/silicon nitride (Si$_3$N$_4$) [10]. The test under reciprocating sliding showed very different results for tests under low humidity (winter conditions) and high humidity (summer conditions). The different wear values under different humidity are due to the formation of different interfacial tribochemical layers. For the steel/silicon carbide couple, the interfacial layers reduced both friction and wear, whereas for the steel/silicon nitride couple the wear coefficient increased but the friction coefficient remained unchanged. The experimental findings underline the important influence of atmospheric humidity on interfacial tribocheneistry and the resulting tribodata.

3.7 Tribomaps

Tribomaps characterize regimes of different tribological processes (or different ranges of friction and wear data) by giving boundaries of operational parameters for these regimes.

For the design of empirical wear mechanism maps, Ashby and co-workers [11] have used normalized values for wear rate, load, and sliding velocity defined by: $W^* = W/A_0$, $F_N^* = F_N/A_0 \cdot H_0$, $v^* = v \cdot r_0/a$, where A_n is the nominal (apparent) contact area of the

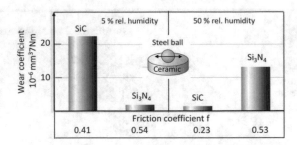

Fig. 3.24 Example of the strong effect of atmospheric humidity on wear of ceramic/steel pairs

wearing surface, H_0 is the room-temperature hardness, a is the thermal diffusivity, r_0 is the radius of the circular nominal contact area, W* is the volume lost per unit area of surface, per unit distance slid; F_N* is the nominal pressure divided by the surface hardness, and v* is the sliding velocity divided by the velocity of heat flow. In analyzing the results of wear tests published in the literature, the different dominating wear mechanisms were classified into the following categories:

- Seizure
- Melt-dominated wear
- Oxidation-dominated wear (mild and severe oxidational wear)
- Plasticity-dominated wear (including delamination wear).

For each wear regime, the boundaries of the operational parameters F_N and v were estimated; the resulting wear map is shown in Fig. 3.25.

3.8   Performance Limits of Tribological Systems

The performance limits of tribosystems are characterized by Transition Diagrams. They are graphical representations of operational parameters separating the region of normal operation from the region of failure. Transition diagrams were developed by the International Research Group (IRG-OECD) [12]. The methodology is illustrated in Fig. 3.26 for the example of lubricated sliding Hertzian steel point contacts (4-ballsystem).

Step 1: Tribological tests should be run with a defined lubricated Hertzian contact system (counterformal point or line contact) under constant operating conditions of sliding velocity v, bulk oil temperature T, sliding distance s, and test duration t, with a stepwise increased load, F_N. The tests may be run either with new specimens for each load step (that is, a "no run-in procedure" where $t < 10$ s), or continuously with one specimens set (that is, a "run-in procedure"). The friction coefficient, f, and wear volume W are then measured and the wear coefficient, $k = W/(F_N \cdot s)$, is calculated.

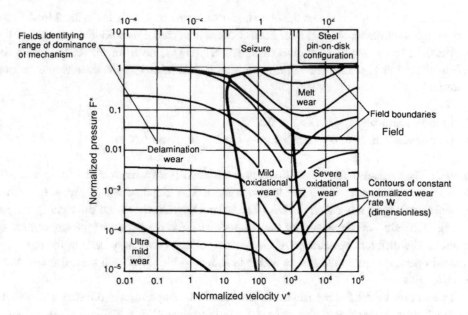

Fig. 3.25 Empirical wear mechanism map for steel (pin-on-disk configuration)

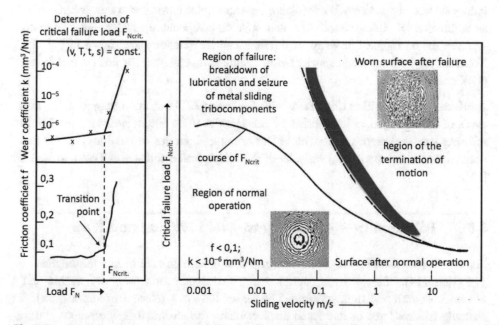

Fig. 3.26 The determination of critical failure loads for sliding lubricated concentrated contacts and the resulting transition diagram (critical failure load as a function of sliding velocity)

Step 2: For a constant set of the operational variables (v, T, t, s), the critical load F_{Ncrit} for the transition from the region of normal operation (partial elastohydrodynamic, EHD, lubrication) to the region of failure (incipient scuffing) is determined. As shown on the left hand side of Fig. 3.26, the failure transition is determined by the following transition criteria:

- friction coefficient $f < 0.1 \rightarrow f > 0.3$,
- wear coefficient $k < 10^{-6}$ mm^3/N m $\rightarrow k > 5 \times 10^{-6}$ mm^3/N m.

Step 3: The critical failure load, F_{Ncrit} is determined for a variation of sliding velocity, v (the other parameters T, t, s are held constant). The graph of F_{Ncrit} versus the corresponding values of v is called the IRG Transition Diagram shown on the right hand side of Fig. 3.26. Beyond the course of the failure load F_N is the region of the termination of motion. The different appearance of the contact surface (stationary ball) in the region of normal operation and in the failure region is shown in Fig. 3.26 with optical interference micrographs.

In addition to the 2-dimensional failure transition diagram, a 3-dimensional "failure surface" representing the critical set of operating variables $\{F_N, v, T\}_{critical}$ at the transition from the region of normal operation to the region of failure can be obtained [12]. The failure surface characterizes the load-carrying capacity of lubricated concentrated contacts as a function of sliding velocity, v, and bulk oil temperature, T. Examples of failure surfaces for tribological, systems of different contact geometry are shown in Fig. 3.27. Note that the load-carrying capacity before failure is much larger for line contact than for point contact.

Application of transition diagrams. With the use of IRG transition diagrams, the influences of materials properties, lubricant characteristics (for example, viscosity or chemical additives), or environmental conditions (for example, inert gas or humidity) on the functional limits of the operational variables of sliding lubricated contacts can be characterized [13].

3.9   Tribotronics—Technology to Avoid Friction and Wear

The term tribotronics applies to the integration of electronic control in a tribological.

System [14]. The electronic control network consists of a sensor, an actuator, and a control unit with real time software. The sensor detects a relevant tribomeasurand, for example friction force or interfacial displacement. The electrical sensor output is transferred via the controller unit to an actuator who act on the tribocontact in order to optimize the tribological system behavior. The concept of tribotronics is illustrated in Fig. 3.28.

A comparison between a classical tribological system and a tribotronic system is made in Fig. 3.29.

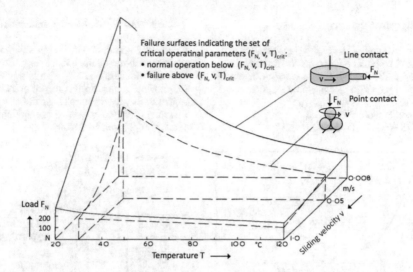

Fig. 3.27 Failure surfaces indicating the transition from the region of normal operation to failure forsliding lubricated point and line contacts

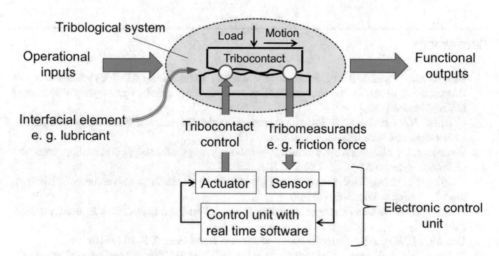

Fig. 3.28 Principle of tribotronic systems

Tribological System: Sliding bearing

Principle: Support of moving parts (load F_N, speed v) with friction. The friction force F_N is influenced by the interfacial element (lubricant) and characterized by the Stribeck curve.
I   Solid friction $f \approx 0,1...>1$,
II Mixed friction $f \approx 0,01...0,1$
III Fluid friction $f \approx 0,001...0,01$

Tribotronic System: Magnetic bearing

Principle: Support of moving parts without physical contact by the electromagnetic levitation of actuators. The air gap s between the moving parts is kept constant at all loads and speeds with a controller-actuator unit. Friction is very low and wear is eliminated. Friction coefficient $f < 0,0001$

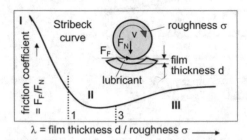

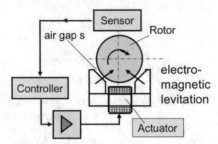

Fig. 3.29 Examples of tribological and tribotronic systems

References

1. Britain, G.: Department of Education and Science: Lubrication (Tribology) Education and Research—A Report on the Present Position and Industry's Needs. Her Majesty's Stationery Office, London (1966)
2. Czichos, H.: Tribology—A Systems Approach to the Science and Technology of Friction, Lubrication and Wear. Elsevier, Amsterdam (1978)
3. Holmberg, K., Ali Erdemir, A.: Influence of tribology on global energy consumption, costs and emissions. Friction **5**(3), 263–284 (2017)
4. Czichos, H., Habig, K.-H. (eds.): Tribologie-Handbuch (Including an Overview of Tribolog), 5th edn. Springer, Wiesbaden (2020)
5. Buckley, D.H.: Surface Effects in Adhesion, Friction, Wear, and Lubrication. Elsevier, Amsterdam (1981)
6. Urbakh, M., Meyer, E.: The renaissance of friction. Nat. Mater. **9**, 8–10 (2010)
7. Czichos, H.: Influence of asperity contact conditions on the failure of sliding elastohydrodynamic contacts. Wear **41**, 1–14 (1977)
8. Stribeck, R.: The principal properties of sliding and rolling bearings. VDI Z. **46**, 1902, 1341, 1432, 1463 (in German)
9. Cowan, R.S., Winer, W.O.: In: Handbuch, T., Czichos, H., Habig, K.-H. (eds.) Machinery Diagnostics: Fundamentals and Tribosystem Applications. Springer, Wiesbaden (2020)
10. Czichos, H., Woydt, M.: Tribological Testing and Presentation of Data, ASM Handbook of Friction, Lubrication and Wear Technology. ASM International (2017)
11. Lim, S.C., Ashby, M.F., Bruton, J.H.: Wear-Rate Transitions and Their Relationship to Wear Mechanisms. Acta Metall. **35**, 1343–1348 (1987)

12. Czichos, H.: Failure criteria in thin film lubrication: the concept of a failure surface. Tribol. Int. **7**, 14 (1974)
13. Lossie, C.M., Mens, J.W.M., de Gee, A.W.J.: Practical applications of the IRG transition diagram technique. Wear **129**, 173–182 (1989)
14. Glavatskih, S., Hoglund, E.: Tribotronics—Towards active tribology. Tribol. Int. **41**, 934–939 (2008)

Mechatronic Systems

4

The term *Mechatronics* was coined in the 1960s in Japan and is now globally acknowl-
edged. Its definition has been expressed as follows [1]

- *Mechatronics is an interdisciplinary field of engineering, including mechanics, electronics, controls, and computer engineering.*

A collection of prominent examples of mechatronic systems from the macroscale to the
nanoscale is given in Fig. 4.1.

4.1   Principles of Mechatronic Systems

Mechatronic systems are characterized by their *structure* and *function*, according to the
General Systems Theory. Mechatronic systems have a basic mechanical structure with
multidisciplinary mechatronic modules and transform via the systems' structure operational inputs (energy, materials, information) into outputs needed to perform a technical
function. The general scheme of mechatronic systems is shown in Fig. 4.2 [2].

The functional performance of mechatronic systems is controlled with a combination
of actuators and sensors, and appropriate algorithms. The general block diagram for the
control loop of mechatronic systems is shown in Fig. 4.3.

The major components of a mechatronic system control unit include a sensor, a controller and an actuator as final control element. The controller monitors the controlled
process variable x, and compares it with the reference, the set point x_{SP}. The difference
between actual and desired value of the process variable, called the controller error $e(t) =
x_{SP} - x$, is applied as feedback to generate a control action to bring the controlled process
variable to the same value as the set point. The diagram of Fig. 4.3 shows a closed loop

© The Author(s), under exclusive license to Springer Nature Switzerland AG 2022   73
H. Czichos, *Introduction to Systems Thinking and Interdisciplinary Engineering*,
Synthesis Lectures on Engineering, Science, and Technology,
https://doi.org/10.1007/978-3-031-18239-6_4

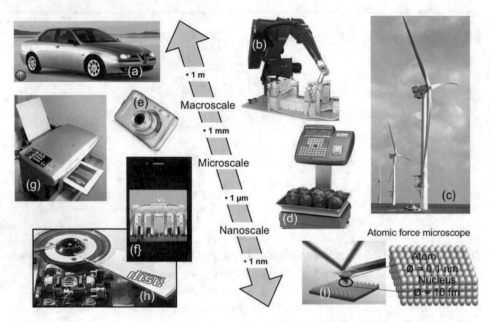

Fig. 4.1 Prominent mechatronic systems: **a** automobile, **b** robot, **c** wind energy converter, **d** weighing machine, **e** digital camera, **f** smartphone, **g** inkjet printer, **h** CD player, **i** atomic force microscope

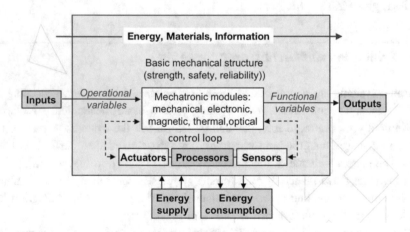

Fig. 4.2 General scheme of mechatronic systems

system based on feedback. A closed loop controller has a feedback loop which ensures the controller exerts a control action to give a process output the same value as the "reference input" or "set point". To implement intermediate value control, the sensor has to measure the full range of the process variable.

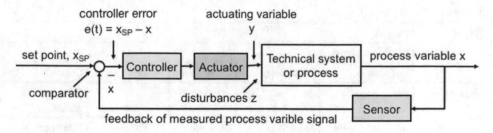

Fig. 4.3 Block diagram for the control of mechatronic systems

4.2   Sensors: The Key to Mechatronic Systems

Sensors have a central role in mechatronic systems as illustrated in Fig. 4.4. They determine functional and structural parameters of interest and transform them into electrical signals that can be displayed, stored or further processed for applications of mechatronic systems [3].

Sensor principles which can be utilized to transform a non-electrical physical parameter of technical systems in an electrical signal are compiled in Table 4.1.

Characteristics of Sensors
A sensor can be illustrated by a block diagram and described by the sensor function $y = f(x)$ which relates the output y to the input x, see Fig. 4.5. If there is a linear input–output relation, the sensitivity ε of the sensor is defined as $\varepsilon = \Delta y/\Delta x$, and its reciprocal is the

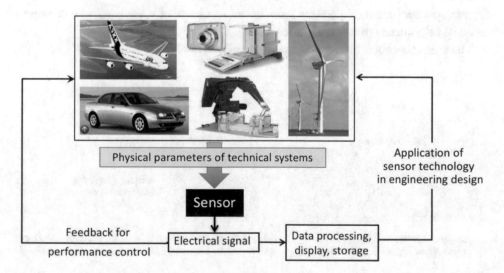

Fig. 4.4 Application of sensors in mechatronic systems

Table 4.1 Sensor selection matrix

Input	Output				
	Electrical signal				
	Resistance	Inductance	Capacitance	Voltage	Current
Strain $\varepsilon =$ $\Delta l/l$	Strain gage				Fiber optical sensor
Position: length l, distance s angle	Resistive sensor, magneto-resistive sensor	Inductive position sensor	Capacitive position sensor	Hall sensor, triangulation sensor	Eddy current sensor, opto-electric sensor
Velocity v = ds/dt	Magneto-resistive angle sensor	Inductive angle sensor		Magnetopole sensor	Optoelectronic revolution sensor
Acceleration a = dv/dt = d^2s/dt^2	Seismic sensors with dynamic mass-damper-spring elements, sensing of mass displacement with electronic position sensor (e.g. Hall sensor)				
Force F torque F · l	Piezo-resitive sensor, strain gauge sensor	Magneto-elastic sensor		Piezoelectric sensor	Sensors with elastic elements and strain or displacement sensor
Pressure p = force/area	Piezo-resitive sensor				
Temperature T	NTC/PTC resistor			Thermo-couple	Optoelectronic pyrometer

input/output coefficient c = $\Delta x/\Delta y$. From the measurement of y, the parameter of interest x can be determined through the relation x = c · y.

The sensor line must be determined by the calibration of the sensor, see Fig. 4.6.

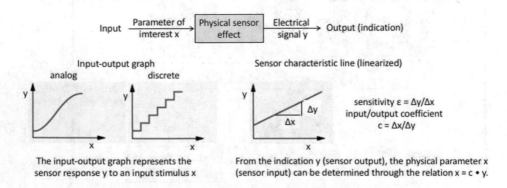

Fig. 4.5 Block diagram of a sensor and the graphical representation of its function

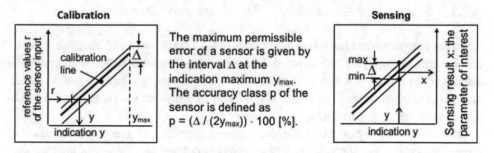

The maximum permissible error of a sensor is given by the interval Δ at the indication maximum y_{max}. The accuracy class p of the sensor is defined as $p = (\Delta / (2y_{max})) \cdot 100$ [%].

Fig. 4.6 Calibration of a sensor and the definition of the accuracy class of a sensor

A calibration diagram represents the relation between indications of a sensor and a set of reference values of the measurand. At the maximum indication value y_{max}, the width Δ of the strip of the calibration diagram is the range of the maximum permissible errors. The accuracy class p of a sensor is defined as $p = (\Delta/(2y_{max})) \times 100$ (%).

Measuring Chain for Sensing

For the application of sensors, a *measuring chain* has to be established. It consists basically of the sensor and units for signal-processing, data acquisition and display. Figure 4.7 illustrates a measuring chain for sensing in its simplest form with its basic terms and formulas.

- A measuring chain consists in the simplest case of
  - (1) the sensor (sensitivity ε_1, accuracy class p_1)
  - (2) a signal-processing unit (sensitivity ε_2, accuracy class p_2)
  - (3) a display (sensitivity ε_3, accuracy class p_3)
- The sensor output can be displayed or used as control signal for the physical parameter, or as signal for sesnsor-based applications (apps), or as actuator input.
- Sensitivity of a measuring chain (stationary operation): $\varepsilon_{measuring\,chain} = \varepsilon_1 \bullet \varepsilon_2 \bullet \varepsilon_3$
- Uncertainty budget of a measuring chain: $p_{measuring\,chain} = \sqrt{(p_1^2 + p_2^2 + p_3^2)}$

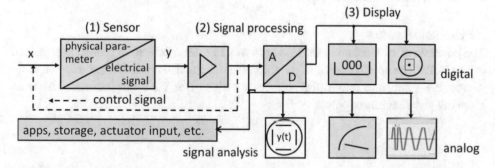

Fig. 4.7 The measuring chain for sensing and its basic features

4.2.1   Sensors for Dimensional Metrology and Kinematics

These sensors transform the geometric input quantity length into an electrical output signal that can be displayed, stored or used for further applications. Kinematics deals with motion of an object in a frame of reference. In the commonly used Cartesian coordinate system, the position of an object is described by the value set $\{x, y, z\}$ of the three rectangular axes x, y, z in space.

Any motion is composed of linear displacement (translation) or/and rotation along or around the three axes. The velocity v of an object is defined as variation of position with time, e.g. $v_x = dx/dt$. Acceleration a is defined as variation of velocity with time, e.g. $a_x = dv_x/dt = d^2x/dt^2$.

Sensing of kinematic parameters begins by describing the geometry of the system and declaring the initial values of the position set $\{x, y, z\}$. Sensors for kinematics are classified as

- position sensors,
- velocity sensor,
- acceleration sensors.

Position Sensing
Position sensors are based on different physical principles to transform a geometric input, i.e. the position of an object to be determined, into electrical, magneto-electric or optoelectronic output signals, see Fig. 4.8.

Velocity Sensing
The velocity v of a moving object is the rate of change of is positon s with time t, $v = ds/dt$. A measuring chain for determining the velocity of an object can be obtained by using the electrical output signal of a positon sensor as input to a differentiator which gives as output the velocity signal. The velocity measuring chain and the principle of an electronic differentiator are explained in Fig. 4.9.

Acceleration Sensing
The principle of an accelerometer (also known as seismic acceleration sensor) is explained in Fig. 4.10. The acceleration a of an object (sensor input) is transformed with a spring-damper-mass structure into a displacement x which gives via a displacement sensor an electrical signal as sensor output.

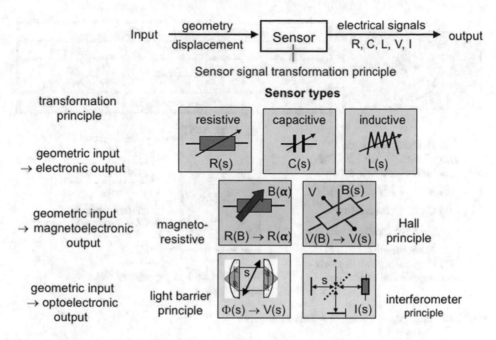

Fig. 4.8 Physical principles for position and displacement sensors

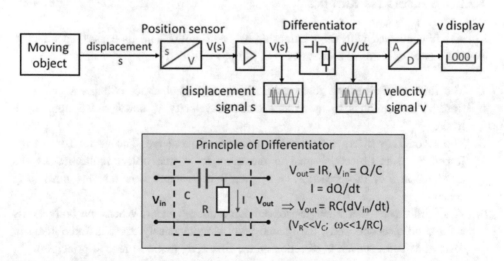

Fig. 4.9 Principle of a velocity measuring chain

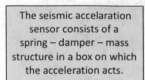

The seismic accelaration sensor consists of a spring – damper – mass structure in a box on which the acceleration acts.

Elements of the seimic acceleration sensor		
Spring k	Damper d	Mass m
$\rightarrow{F_k}$ y	$\rightarrow{F_d}$ $dy/dt = v$	$\rightarrow{F_m}$ $d^2y/dt^2 = a$
$F_k = k \cdot y$	$F_d = d \cdot v = d \cdot dy/dt$	$F_m = m \cdot d^2y/dt^2$

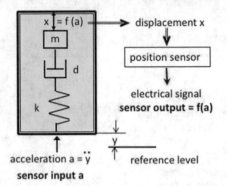

The balance of force components $\sum F = 0$ gives the sensor differential equation

$$m\ddot{x} + d\dot{x} + k\,x = F(t) = m\ddot{y}$$

If $k \gg m$, d it follows that

$$x = F(t) \approx m/k \cdot \ddot{y}$$

Measurement of the displacement x with a position sensor yields the acceleration a.

Fig. 4.10 Principle of a seismic acceleration sensor

4.2.2 Sensors for Kinetics

Kinetics is concerned with the relationship between motion and its causes, specified in Newton's laws of classical mechanics:

I. The first law defines the force qualitatively: an object either remains at rest (stationary) or continues to move at a constant velocity v, unless acted upon by a force.
II. The second law offers a quantitative measure of the force. The vector sum of the forces F on an object is equal to the mass m of that object multiplied by the acceleration a of the object: $F = m \cdot a$. (It is assumed here that the mass m is constant.)
III. The third asserts that a single isolated force doesn't exist. When one body exerts a force on a second body, the second body simultaneously exerts a force equal in magnitude and opposite in direction on the first body (actio = reactio principle).

Force Sensors
Force is a vector with a direction and a magnitude, it can only be determined by the response to its action. The design of force sensors utilizes the principles shown in Fig. 4.11.

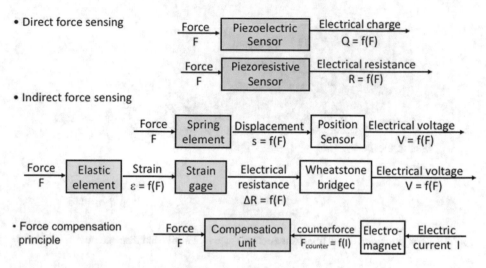

- Direct force sensing
- Indirect force sensing
- Force compensation principle

Fig. 4.11 The principles of force sensors

Direct Force Sensors

Direct force sensing is realized with piezo-electric sensors. They transform the input of a force directly in an output of electrical charge or electrical resistance as described in Fig. 4.12.

Piezoelectric sensors exploit the piezoelectric effect. The action of a force on a piezoelctric crystal results in charge polarization across the crystal, for example quarz, zinc oxide or the polymer polyvinylfluoride (PVDV).
- Sensing circuit:
The piezosensor crystal is effectively a capacitor that generates a charge Q proportional to the force F deforming the crystal. The sensor is mechanically very stiff. It allows only dynamic measurements, the opeartnig frequency should be above 1 Hz.

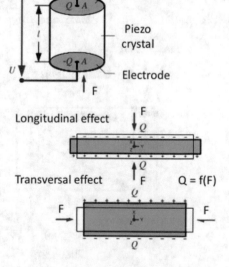

Longitudinal effect

Transversal effect

$Q = f(F)$

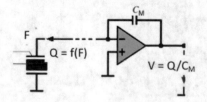

Fig. 4.12 The principle of piezoelectric force sensors

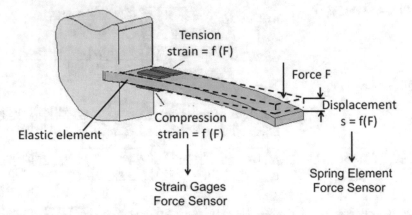

Fig. 4.13   The principles of the transformation of force into strain and displacement

Indirect Force Sensors

The principles for indirect force sensing are illustrated in Fig. 4.13. A force acting on an elastic element causes displacement and strain. This lead to two types of force sensors.

- Spring element force sensors, exemplified as load cell in Fig. 4.14, and
- Strain gages force sensors, exemplified in Fig. 4.15.

- **Functional principle of force-measuring load cell**

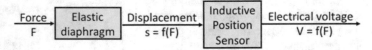

- **Sensor design**

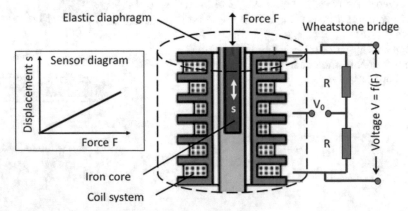

Fig. 4.14   The principle of a load cell, designed with an elastic diaphragm and an inductive position sensor

Principle of force sensors with strain gages

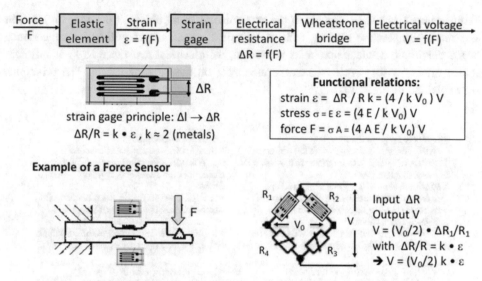

Functional relations:
strain $\varepsilon = \Delta R / R\,k = (4 / k\,V_0)\,V$
stress $\sigma = E\,\varepsilon = (4\,E / k\,V_0)\,V$
force $F = \sigma\,A = (4\,A\,E / k\,V_0)\,V$

Example of a Force Sensor

Input ΔR
Output V
$V = (V_0/2) \bullet \Delta R_1/R_1$
with $\Delta R/R = k \bullet \varepsilon$
$\rightarrow V = (V_0/2)\,k \bullet \varepsilon$

Fig. 4.15 The principle of a force, designed with strain gages

Force Compensation Principle

The force F to be measured is compensated by a force F_C, created by an electromagnet as function of current I until the lever arm system is in balance, i.e. x = 0. At the balance position, F_C is equal to F, and the current I is a measure of F, see Fig. 4.16.

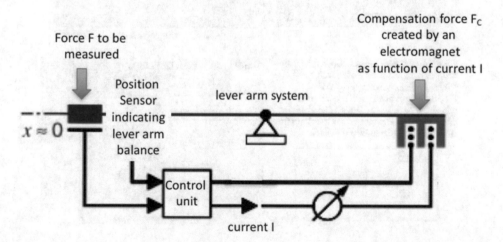

Fig. 4.16 The force compensation principle

4.2.3   Temperature Sensors

Temperature T is a state variable of the heat contents of an object, it depends on location and time. Temperature sensors with an electrical signal output utilize the dependence of the electronic conductance or its reciprocal, the electrical resistance R(T) as physical sensor effect, or they are based as thermocouple on the Seebeck-Effect. The principles are explained in Fig. 4.17.

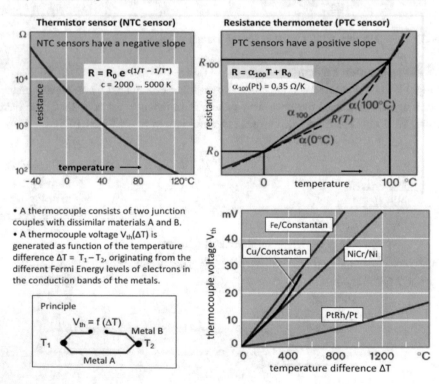

Thermistor temperature sensor
• Semiconductors have a high electrical resistance at low temperatures because there are few electrons in the conduction band.
• Increasing temperature T activates electrons into the conduction band. Thus, the conductance increases and the resistance R decreases with increasing temperature.
• Thermistor temperature sensors have a negative slope in their R-T diagram and are called NTC sensors.

Resistance thermometer
• The electrical resistance R of conductors results from the scattering of the conduction electrons at the atoms or molecules of the solid lattice.
• Increasing temperature T intensifies lattice vibrations. Thus, the electrical resistance increases with increasing temperature.
• Resistance thermometer have a positive slope in their R-T diagram and are called PTC sensors.

Thermistor sensor (NTC sensor)

NTC sensors have a negative slope

$$R = R_0\, e^{\,c(1/T - 1/T^*)}$$
$$c = 2000 \dots 5000\ K$$

Resistance thermometer (PTC sensor)

PTC sensors have a positive slope

$$R = \alpha_{100}T + R_0$$
$$\alpha_{100}(Pt) = 0{,}35\ \Omega/K$$

• A thermocouple consists of two junction couples with dissimilar materials A and B.
• A thermocouple voltage $V_{th}(\Delta T)$ is generated as function of the temperature difference $\Delta T = T_1 - T_2$, originating from the different Fermi Energy levels of electrons in the conduction bands of the metals.

Principle

$$V_{th} = f\,(\Delta T)$$

Fig. 4.17 Principles of temperature sensors

4.2.4   Embedded Sensors

Embedded sensors detect "in situ" loads on the components of technical systems and transform them into electrical signal which can be used to control the structural integrity or the functional behavior of the component [4]. They are tied in the structural components of technical systems either by affixing them with a glue (e.g. strain gages) or encase them in a component (e.g. fiber optical sensor). By adding an actuator network and a related control and drive system, the structure can be improved to a so-called "smart technical system". The application of embedded sensors is exemplified in Fig. 4.18.

Embedded Sensor Technology for Infrastructure Safety
"Infrastructure" is the set of fundamental facilities necessary for society and economy to function. Classical examples are supply and disposal facilities for energy and water as well as communication and transport networks. Infrastructure includes the physical components of interrelated systems providing commodities and services essential to enable, sustain, or enhance societal living conditions and maintain the surrounding environment. To improve the functionality and safety of public infrastructure, mechatronics—especially sensor technologies—have been utilized. A prominent example is Berlin's completely new Central Station, Fig. 4.19 The German Television informed the public about the new application of advanced technology for infrastructure safety as follows:

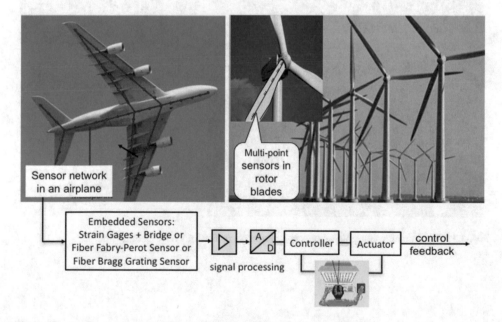

Fig. 4.18 Application examples for embedded sensors

At Berlin's new Central Station - one of the most complicated and complex buildings in Europe - a total of more than 200 sensors were embedded at all statically relevant structural components of the station. The sensors measure strains, deformations and settlements using different physical principles such as electrical, optical and mechanical transducers. The strain sensors for example, are attached directly to the building structure, where even the finest deformation can be registered by changing their electrical resistance of the sensor. Over 3000 meter cables are laid through cavities to ensure the online monitoring of the entire construction. The collected signals are converted into physical units and processed further. A special developed computer program evaluates the condition data. Thus a current status picture of the station at particularly stressed locations can be called up at any time. This new dimension in structural safety has been carried out 2004 for the first time at Berlin's new Central Station.

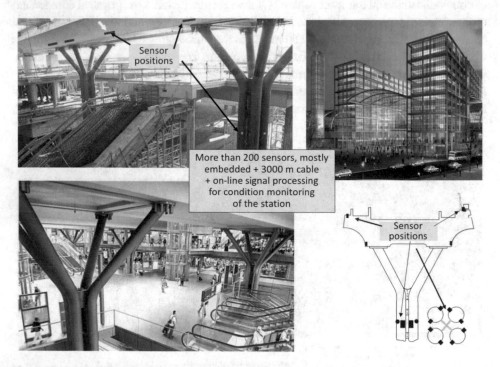

Fig. 4.19   Application example of sensor technology in Berlin's new Central Station

4.2.5 Remote Sensing

Wave propagation sensors for the determination of distance and speed operate with transmitted and reflected waves of different wavelength ranges, categorized as follows:

RADAR (Radio detection and ranging) is an object-detection system that uses radio waves to determine the range, angle, or velocity of objects. A radar system consists of a transmitter producing electromagnetic waves in the radio or microwave domain, a transmitting antenna a receiving antenna (often the same antenna is used for transmitting and receiving) and a receiver and processor to determine properties of the object. Radio waves (pulsed or continuous) from the transmitter reflect off the object and return to the receiver, giving information about the object's location. The velocity v of a moving object, e.g. the speed of a car, can be detected with the Doppler shift.

LIDAR (Light detection and ranging) is a surveying method that measures distance to a target by illuminating that target with a pulsed laser light, and measuring the reflected pulses with a sensor. Differences in laser return times and wavelengths can then be used to measure distances. LIDAR uses Ultraviolet, Visible, or near Infrared light to image objects. It can target a wide range of materials, including non-metallic objects, rocks, and chemical compounds. A narrow laser-beam can map physical features with very high resolution.

IR Sensors are device that forms an image using infrared radiation up to $\lambda = 14$ μm. This sensor technique can also be used for remote-sensing temperature.

Ultrasonic Sensors evaluate targets by interpreting the reflected signals. For example, by measuring the time between sending a signal and receiving an echo, the distance of an object can be calculated.

An overview of the physical principles of wave propagation sensors is given in Fig. 4.20.

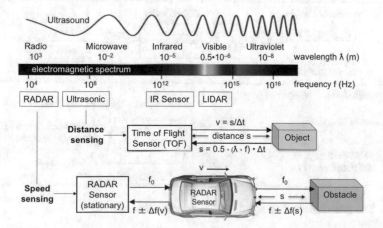

Fig. 4.20 The physics of wave propagation for remote sensing distance and speed

4.3   Actuators in Mechatronic System

An actuator (in short actor) is a component in a mechatronic system that is responsible for the functional output by which the system acts to fulfil its task [5]. An actuator requires a control signal and an electrical, fluidic or thermal power supply.

- Actuators with electric power supply
  - electric actuators, powered by a motor that converts electrical energy into mechanical
  - galvanometer actuators produce a rotary deflection in an electro-magnetic field
  - piezoelectric actuators, based on the inverse piezo-electrical effect
- Actuators with fluidic power supply
  - hydraulic actuators, consisting of a fluid motor
  - pneumatic actuators, converting energy formed by vacuum or compressed air
- Actuators with thermal power supply
  - bimetallic strip that converts a temperature change into mechanical displacement
  - shape-memory alloy, that switches between two crystal structures of different shape.

The principles of actuators and the different types are illustrated in Fig. 4.21.

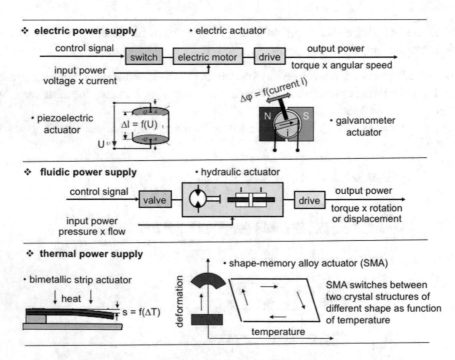

Fig. 4.21 The general principles and the different types of actuators

The ranges of the actuator's functional outputs *force, displacement, and velocity* of the various actuator types are depicted in Figs. 4.22 and 4.23 [5]. In Fig. 4.23 also the output power range is plotted.

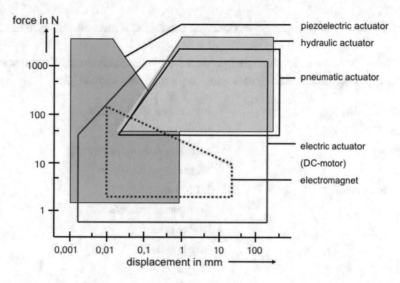

Fig. 4.22   Typical ranges of force and displacement of actuators

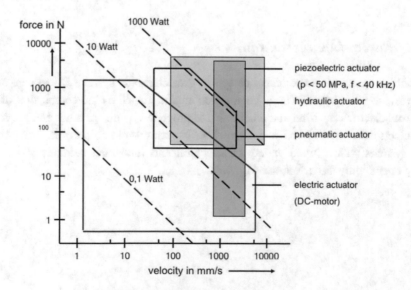

Fig. 4.23   Typical ranges of force and velocity of actuators

4.4 Applied Mechatronics

Mechatronic systems are created to fulfil basic needs of technology, industry and society. Their significance has been expressed as follows:

- Virtually every newly designed engineering product is a mechatronic system [1].

Mechatronic systems are designed and produced by **Interdisciplinary Engineering**. The tasks and types of mechatronic systems are very broad and include [2]:

• Human mobility	– Automobile mechatronics
• Industrial handling	– Robot
• Weighing	– Mechatronic weighing system
• Power supply	– Wind energy converter
• Micro-mechatronics	– MEMS and MOEMS
• Commodities	– Smartphone mechatronics
	– Digital camera
	– Inkjet printer
	– CD player
• Nano-mechatronics	– Atomic force microscopy

4.4.1 Automobile Mechatronics

Automobiles (cars) are a prominent category of mechatronic systems. Most cars are traditionally propelled by an internal combustion engine, fueled by the combustion of fossil fuels. An "electric car" is an automobile that is propelled by one or more electric motors, using electrical energy stored in rechargeable batteries. Automobiles are today mechatronic systems with a broad variety of structural and functional modules and controls usually operated by the driver, see Fig. 4.24.

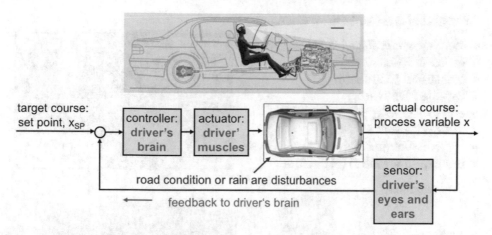

Fig. 4.24 A driving automobile seen as a mechatronic system with the driver acting with his eyes and ears as sensor, with his brain as processor, and with his muscles as actor

Sensors in the Automobile

From a technological point of view, virtually all parameters relevant for the road performance of an automobile, the safety and comfort of car passengers, and environment protection can be determined with appropriate sensors and controlled with mechatronic modules which include actuators and control algorithms. A selection of sensors for functional tasks in an automobile is shown in Fig. 4.25. The sensors are part of a mechatronic control system consisting of sensors, actuators, a bus (bidirectional universal switch) system and a central control and display unit.

Dynamic Stability Control

Driving stability is of outmost importance for all automobiles—be they propelled by an internal combustion engine or by an electric motor, Dynamic stability control (DSC), also referred to as electronic stability program (ESP) is a mechatronic technology that improves a car's driving stability by detecting and reducing loss of traction (skidding). DSC estimates with sensors the direction of the skid, and then applies via actuators the brakes to individual wheels asymmetrically in order to create torque about the vehicle's vertical axis, opposing the skid and bringing the vehicle back in line with the driver's commanded direction (denoted as "target course set point" in Fig. 4.24).

During normal driving, the dynamic stability control works in the background and continuously monitors steering and vehicle direction. The DSC module compares the driver's intended direction x_{SP} (determined through the measured steering wheel angle) to the vehicle's actual direction x (determined through measured lateral acceleration, vehicle rotation, and individual road wheel speeds).

The principle of mechatronic dynamic stability control is explained in Fig. 4.26.

Functional tasks and sensor examples

- steering wheel angle (AMR sensor)
- cam shaft (Hall sensor)
- crank shaft (Hall sensor)
- throttle (Hall sensor)
- anti-lock braking (inductive sensor)
- airbags (seismic sensor)
- tire pressure (micro mechanics sensor)
- temperature control (PTC sensor)

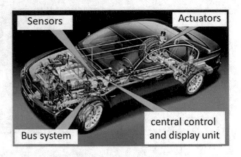

Examples

Hall sensor unit

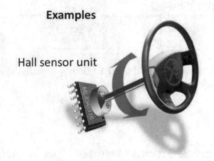

Fig. 4.25 Examples of sensors in an automobile

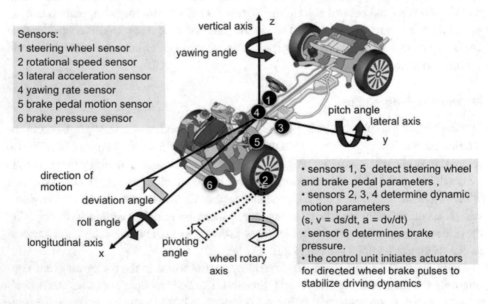

Sensors:
1 steering wheel sensor
2 rotational speed sensor
3 lateral acceleration sensor
4 yawing rate sensor
5 brake pedal motion sensor
6 brake pressure sensor

vertical axis
yawing angle

pitch angle
lateral axis
y

direction of motion
deviation angle
roll angle
longitudinal axis
x
pivoting angle
wheel rotary axis

- sensors 1, 5 detect steering wheel and brake pedal parameters ,
- sensors 2, 3, 4 determine dynamic motion parameters
(s, v = ds/dt, a = dv/dt)
- sensor 6 determines brake pressure.
- the control unit initiates actuators for directed wheel brake pulses to stabilize driving dynamics

Fig. 4.26 Principle of dynamic stability control

Oversteer yaw momentum clockwise	Understeer yaw momentum anticlockwise
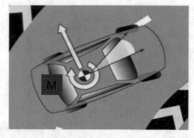	
ESP-correction: brake pulse to the outer front wheel, correcting yaw momentum anticlockwise	ESP-correction: brake pulse to the inner rear wheel, correcting yaw momentum clockwise

Fig. 4.27 The control function of mechatronic dynamic stability control

Driving stability is achieved with individual brake pulses to the outer front wheel to counter oversteer or to the inner rear wheel to counter understeer, see Fig. 4.27.

4.4.2   Industrial Robot

Robots are a prominent category of mechatronic systems and indispensable today for industrial handling and production technologies [6]. They are automated, programmable and capable of movement on two or more axes. Instrumental for all these tasks is the mechatronic combination of mechanics, electronics, informatics and data mining with sensor technology.

An overview of the basic feature of an industrial robot is given in Fig. 4.28 On the left, the general description of the function and the structure is given, according to the general system theory. A sketch of a robot as mechatronic system is shown on the right side of Fig. 4.28. The structure (S) of this mechatronic system is given by the robot components (A), their properties (P) and their motional interrelation (R). The robot function is the transfer (T) of operational inputs (X) into functional outputs (Y). Sensors, processors and

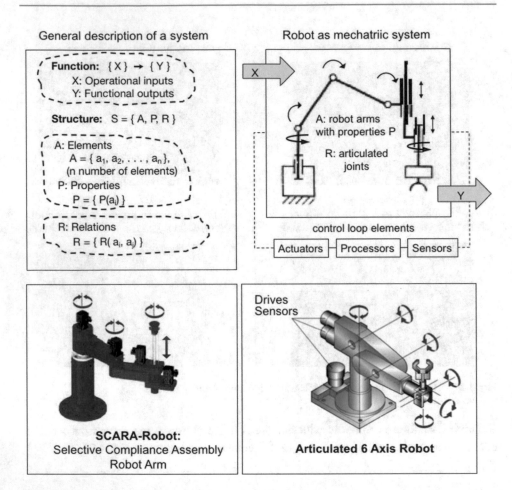

Fig. 4.28 The characterization of robots as mechatronic systems

actuators form a control loop for the performance of the system. The most commonly used robot configurations are Articulated Robots and SCARA robots, illustrated in the lower part of Fig. 4.28.

4.4.3   Mechatronic Weighing System

Weighing is an everyday procedure for all who sell or buy articles, food etc. by weight. In scientific terms, the weight W of a body is equal to the magnitude F_g of the gravitational force on the body: $W = F_g = mg$. It follows that $m = F_g/g$, where g is the gravitational

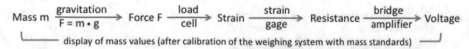

Mass m $\xrightarrow[F = m \cdot g]{\text{gravitation}}$ Force F $\xrightarrow[\text{cell}]{\text{load}}$ Strain $\xrightarrow[\text{gage}]{\text{strain}}$ Resistance $\xrightarrow[\text{amplifier}]{\text{bridge}}$ Voltage

display of mass values (after calibration of the weighing system with mass standards)

Design of the mechatronic weighing system

mass

load cell

strain gage

Wheatstone bridge

U_0

amplifier

U

temperature ΔT compensation

voltage = f (load)

display

000

μP

A D

Fig. 4.29 Principle of a mechatronic weighing system with strain gages

constant. A mechatronic weighing system determines the mass m of an object by measuring the force F_g with a force sensor. A mechatronic weighing system with strain gages is shown in Fig. 4.29.

4.4.4 Wind Energy Converter

Wind is the movement of air across the surface of the Earth, driven by areas of high and low pressure. The atmosphere acts as a thermal engine, absorbing heat at higher temperatures, releasing heat at lower temperatures. Wind energy is the kinetic energy of air in motion $E = 1/2 \, mv^2$. Power is energy per unit time, Wind power in an open air stream is thus proportional to the third power of the wind speed.

A Wind Energy Converter (or Wind Turbine) is a device that converts wind energy, first with a rotor blade into mechanical energy, and then with an induction generator into electrical energy. The function of a Wind Energy Converter and its structural design is illustrated in Fig. 4.30, on the right, the process elements are named.

Wind is the "energy input", necessary for the function of the wind turbine, but this is at the same time a "load" for the wind turbine structure. Loads can be classified as:

- Static loads (horizontal: mean wind and mean currents, vertical: gravity on non-rotating components)

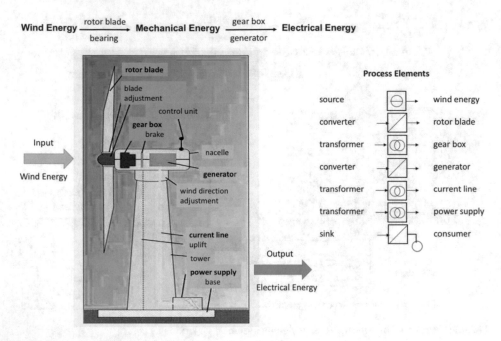

Fig. 4.30 Wind energy converter and basic process elements

- Periodic loads (gravity on rotating components, regular waves, wind shear, tower disturbances, yaw errors)
- Stochastic loads (irregular waves, turbulent wind)
- Transient loads (turbine start-ups and shut- downs, gusts, extreme waves, meandering wakes of neighboring turbines).

Performance monitoring is used for safeguarding the wind turbines. Standard techniques include:

- Vibration analysis with focus on rotating components like bearings and wheels, based on different sensors (position, velocity, accelerometer) and specified for different frequencies and applied to shafts, bearings, gearbox, etc.
- Strain measurements, utilizing electrical or fiber optic, aiming at structural monitoring for lifetime prediction and life cycle management
- Thermography is a remote sensing technique to detect radiation in the long-infrared (IR) range of the electromagnetic spectrum ($\lambda \approx 9 \dots 4 \, \mu\text{m}$). A thermographic camera produces colored images of the radiation of an object. This allows, according to Planck's law, a color-temperature mapping of the object, provided that the emissivity of the object is known. Figure 4.31 shows an application example: a thermogram of a wind turbine propeller, recorded with a helicopter.

Fig. 4.31 Thermogram of a wind turbine propeller, recorded with a helicopter

4.4.5   MEMS and MOEMS

Micro mechatronic systems are miniaturized mechatronic systems created by the methods and techniques of micro-technology. The terms "Micro electro-mechanical systems" (**MEMS**) and "Micro opto-electro-mechanical systems" (**MOEMS**) were coined for miniaturized items with moving parts. The general block diagram of micro mechatronic systems and the scaling requirements for sensors and actuators are shown in Fig. 4.32.

In microscale mechatronics, MEMS and MOEMS are miniaturized actuators. The design of different micro actuators for **MEMS** is illustrated in Fig. 4.33. Conventional

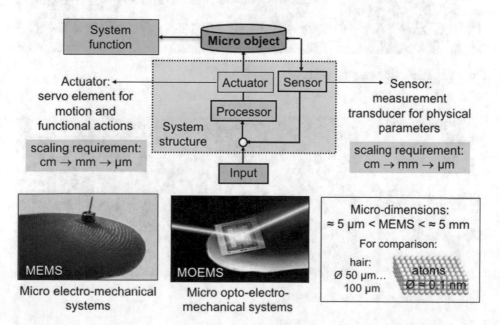

Fig. 4.32 General control-loop block diagram for micro mechatronic systems and the scaling requirements for sensors and actuators

electric motors operate through the interaction between an electric motor's magnetic field and winding currents to generate force or torque. They can be downsized to dimensions of about two millimeters, as shown on the left of Fig. 4.33. Actuators with dimensions in the micrometer range can be achieved with electrostatic motors, an example is shown in the middle of Fig. 4.33. On the right of Fig. 4.33, the characteristic data of piezo actuators are shown. They are based on the "inverse" piezoelectric effect.

Micro actuators for **MOEMS** are exemplified in Fig. 4.34 They utilize switchable electrodes and electrostatic motors for the optical transmission and display of signals and data.

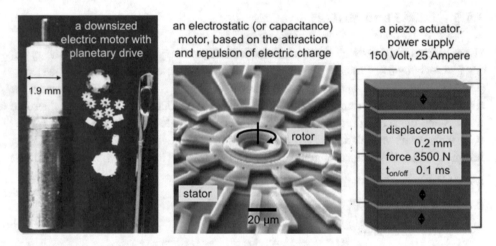

Fig. 4.33   Micro actuators for MEMS

Fig. 4.34   Micro actuators for MOEMs

4.4.6 Smartphone Mechatronics

Smartphones are multi-purpose information and communication devices. They are equipped with various mechatronic modules and sensors. An important sensor principle utilized for the tactile steering of smartphones is the capacitor principle, Fig. 4.35.

A capacitor consists of two conductors separated by a non-conductive region. A voltage V between the two conductors causes electric charges $\pm Q$ and an electric field. The capacitance C is defined by $C = Q/V$. The capacitance changes if the distance between the plates or the overlap of the plates changes.

- The capacity C of a distance sensor is inversely proportional to the input displacement d.
- The capacity of a displacement sensor is directly proportional to the overlap of the capacitor plates.

Capacitive **touchscreen sensors** consist of two transparent conductive traces of indium tin oxide (ITO), separated by a thin insulator, Fig. 4.36. The traces form a grid pattern with x–y coordinates. At the x–y intersections local capacitors arise coupled by electric fields. A finger (or conductive stylus) near the surface of the touchscreen changes the local electric field and reduces the mutual capacitance.

When a finger hits the screen, a tiny electrical charge is transferred to the finger creating a voltage drop at the contact point on the screen. The processing unit detects the x–y location of this voltage drop. This signal is used as steering input for the panel.

The application of a capacitor sensor for **image positioning** in smartphones is illustrated in Fig, 4.37.

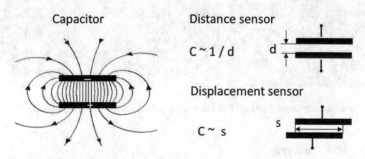

Fig. 4.35 The principles of capacitor sensors

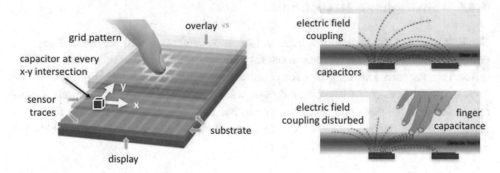

Fig. 4.36   The capacitor sensor in its application in a smartphone

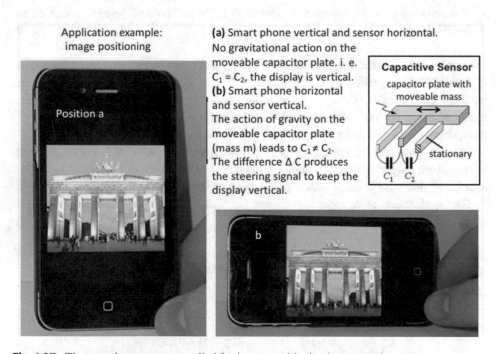

Fig. 4.37   The capacitor sensor as applied for image positioning in a smartphone

4.4.7   Digital Camera

Digital cameras are optical instruments used to record images. A camera captures light photons usually from the visual spectrum for human viewing, but in general could also be from other portions of the electromagnetic spectrum. All cameras use the same basic design: light enters an enclosed box through a lens system and an image is recorded on a light-sensitive medium (sensor). A shutter mechanism controls the length of time that

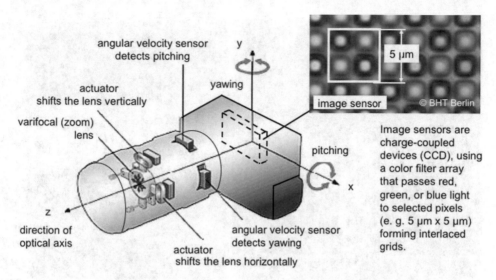

Fig. 4.38 Principle of optical image stabilization with MEMS and MOEMS in digital cameras

light can enter the camera. A digital camera (or digicam) is a camera that encodes digital images and videos digitally and stores them for later reproduction.

Key mechatronic modules are "Micro electro-mechanical systems" (MEMS) and "Micro opto-electro-mechanical systems" (MOEMS), see Sect. 4.5. They are responsible in digital cameras for optical image stabilization. Image stabilization (IS) is a technique that reduce blurring associated with the motion of a camera or other imaging device during exposure. Generally, it compensates for angular movement (yaw and pitch) of the imaging device. Mechatronic image stabilization can also compensate $\pm 1°$ of rotation. The most common actuator is a galvanometer actuator. In combination with strong permanent magnets, two coils are used to drive a platform both vertically and horizontally. As this system inherently creates a strong magnetic field, Hall positions sensors can be used to detect pitching and yawing. The actuators and position sensors can be tightly integrated in a small package, and work together in optical image stabilization, see Fig. 4.38.

4.4.8 Inkjet Printer

Micro piezo actuators are the basic mechatronic modules in inkjet printers. They are a type of printing that recreates a digital image by propelling droplets of ink onto paper with the drop-on-demand technique, DOD. Piezo printers use an actuator nozzle array with a miniaturized ink-filled chamber behind each nozzle. The printing principle is explained

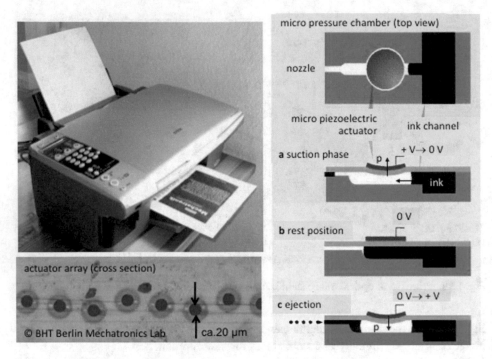

Fig. 4.39 The principle of inkjet printing with micro piezo actuators

in Fig. 4.39. In the suction phase (a), ink is taken into a chamber. In the rest phase (b), the chamber is filled with ink. When a voltage is applied, in the ejection phase (c), the piezoelectric actuator changes shape, generating a pressure pulse in the fluid, which forces a droplet of ink from the nozzle. A DOD process uses software with an algorithm that directs the heads to apply between zero and eight droplets of ink per dot, only where needed.

4.4.9 CD Player

A CD player is a mechatronic device that plays audio compact discs which are a digital optical disc data storage format. A CD is made from 1.2 mm thick, polycarbonate plastic and weighs about 20 g. CD data is represented as tiny indentations ("pits") encoded in a spiral track molded into the polycarbonate layer. The areas between pits are known as "lands".

The dimensions of a CD and a micrograph of the data track are shown in Fig. 4.40.

A CD is read by focusing a 780 nm wavelength (near infrared) semiconductor LASER housed within the CD player, through the bottom of the polycarbonate layer. The change

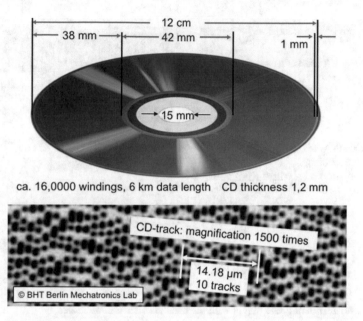

Fig. 4.40 A CD and a micrograph of the data track

in height between pits and lands results in a difference in the way the light is reflected. A change from pit to land or land to pit causes interference of a reflected LASER beam and indicates a bit one, while no change indicates a series of zero bits. Figure 4.41 shows the principle of data retrieval.

When a CD is loaded into the CD player, the data is read out according to the data retrieval principle explained in Fig. 4.41. An electric motor spins the disc with a variable

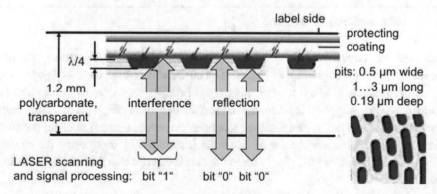

Fig. 4.41 The principle of the retrieval of data stored on a CD

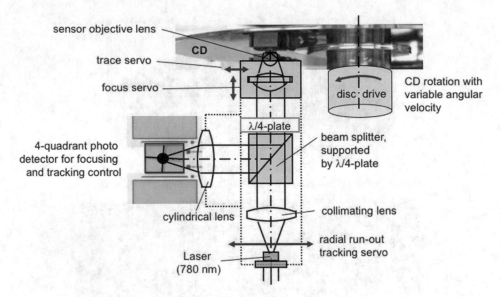

Fig. 4.42 The design of a CD player with the basic mechatronic modules

rotary speed ω to obtain at all positions r a constant reading velocity $v = \omega \cdot r$. The tracking control is done by analogue servo amplifiers and then the high frequency analogue signal read from the disc is digitized, processed and decoded into analogue audio signals. In addition, digital control data is used by the player to position the playback mechanism on the correct track, do the skip and seek functions and display track.

An overview of the mechatronic modules needed for the functional performance of a CD player is given in Fig. 4.42.

4.5 Nano-mechatronics

A mechatronic device which operates at the nanoscale is the **Scanning Tunneling Microscopy** (STM), explained in Fig. 4.43. The STM is based on the physical effect of quantum tunneling. When a conducting tip is brought very near to the surface to be examined, a voltage applied between tip and surface can allow electrons to tunnel through the vacuum between them. The resulting tunneling current is a function of tip position. A surface topography imaging on the atomic scale is acquired by monitoring the current as the tip's position scans with a piezo actuator across the surface, and is displayed in image form.

An **Atomic Force Microscope** (AFM) is an instrument for dimensional metrology of surface topography on the atomic scale. It detects repulsive forces on the nanometer scale by "feeling" or "touching" a surface with a mechanical probe. The design of an AFM is

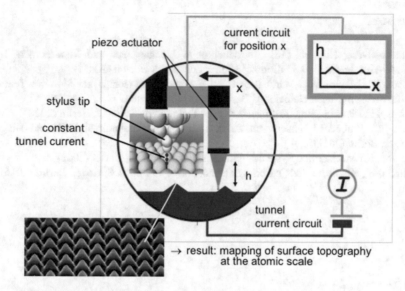

Fig. 4.43 Principle of the scanning tunneling microscope (Nobel prize in Physics 1986 for the inventors Binning and Rohrer)

illustrated in Fig. 4.44. A sharp tip with a radius of approximately 5–20 nm is mounted on a very soft cantilever with a spring constant in the range of 1 μmN/μm. The tip is scanned over the surface by a xyz piezo-actuator with a resolution of less than a nanometer. The displacement range is in the order of 10 μm in the z direction, and up to 150 μm in the lateral x- and y-directions.

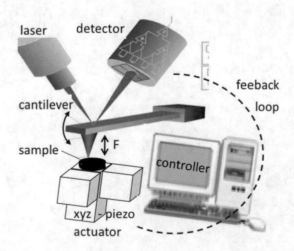

Fig. 4.44 Principle of an atomic force microscope

References

1. Alciatore, D.G., Histand, M.B.: Introduction to Mechatronics and Measurement Systems. McGraw-Hill, International Edition of Higher Education, Boston (2003)
2. Czichos, H.: Mechatronik—Grundlagen und Anwendungen technischer Systeme. Textbook on Mechatronics. Springer, Heidelberg, Berlin (2019) (in German)
3. Czichos, H.: Measurement, Testing ads Sensor Technology. Springer, Berlin (2018)
4. Daum, W.: Embedded sensors. In: Czichos, H. (ed.) Handbook of Technical Diagnostics. Springer, Berlin (2011)
5. Isermann, R.: Mechatronic Systems—Fundamentals. Springer, Berlin (2003)
6. Siciliano, B., Khatib, O. (eds.): Springer Handbook of Robotics. Springer, Berlin (2016)

Cyber-Physical Systems and Industry 4.0

Today's technology developed—based on findings in the natural sciences, new materials and engineering innovations—in four major phases from the middle of the eighteenth century onwards:

- **Period 1**: Industrial Revolution through the development of mechanical technologies with the help from steam engines (James Watt, 1769) and mechanical automation technology (e.g. Loom, 1785)
  - **Mechanization**

- **Period 2**: Electro-mechanics by combining mechanical and electrical technologies (electric motor, generator, Siemens 1866)
  - **Electrification**

- **Period 3**: Systems engineering combination of electro-mechanics with electronics, computer engineering, information technology and digitalization (1960s)
  - **Mechatronics**

- **Period 4**: Combination of Mechatronics and the Internet of things (from 2000 on)
  - **Cyber-physical systems (CPS)**,

  - **Industry 4.0**.

The Technological Concept of Cyber-Physical Systems
The term *Cyber-physical Systems (CPS)* is defined by the German Academy of the Technical Sciences ACATECH, (www.achatech.de) as follows:

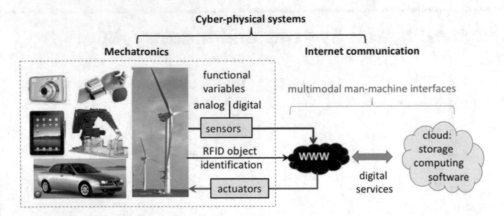

Fig. 5.1 The concept of cyber-physical systems

- *Cyber-physical systems are characterized by a combination of real (physical) objects and processes with information-processing (virtual) objects and processes via open, partly global and at all times connectable information networks.*

Cyber-physical systems merge mechatronics with cybernetics, design and process science and communication. The Internet of things (IoT) is the inter-networking of physical devices, embedded with electronics, sensors, actuators, software, and network connectivity which enable these objects to collect and exchange data.

CPS apply RFID (radio-frequency identification) techniques to identify objects of interest and use sensors that collect physical data and act—with digital networks and actuators—on production, logistics and engineering processes, whereby they utilize multimodal man–machine interfaces. The general scheme of cyber-physical systems consisting of the combination of mechatronics with an internet combination is shown in Fig. 5.1.

The most critical success factor for the realization and introduction of cyber-physical production systems is data, information and communication security. CPS networking creates security threats for industrial production with potential implications for functional safety aspects. Therefore, appropriate security architectures, protective measures and measures for the application of CPS and validation methods are required.

CPSs are of relevance in such diverse industries as aerospace, automotive, energy, healthcare, manufacturing, infrastructure, consumer electronics, and communications [1]. An evolving cyber-physical systems landscape is anticipated for almost all areas of technology with the following examples [2]:

	Present	Future
Energy	Central generation, supervisory control and data acquisition systems for transmission and distribution	Systems for more efficient, safe and secure generation, transmission, and distribution of electric power, integrated through smart grids. Smart ("net-zero energy") buildings for energy savings
Manufacturing	Computer controlled machine tools and equipment. Robots performing repetitive tasks, fenced off from people	Smarter, more connected processes for agile and efficient production. Manufacturing robotics that work safely with people in shared spaces. Computer-guided printing or casting of composites
Materials	Predominantly conventional passive materials and structures	Emerging materials such as carbon fiber and polymers offer the potential to combine capability for electrical and/or optical functionality with important physical properties (strength, durability, disposability)
Transportation and mobility	Vehicle-based safety systems, traction, brake ad stability control, GPS guidance	Vehicle-to-vehicle communication for enhanced safety and convenience, drive-by-wire autonomous vehicles. Next generation air transportation systems
Medical care and health	Pacemakers, infusion pumps, medical delivery devices connected to the patient for life-critical functions	Life-supporting micro-devices embedded in the human body. Wireless connectivity enabling body area sensor nets. Wearable sensors and benignly implantable devices. Configurable personalized medical devices

The Technological Concept of Industry 4.0

The term Industry 4.0 denotes a multidisciplinary technology, which evolved in the beginning of the twenty-first century, to apply the concept of cyber-physical systems (CPS) to industry [3]. Central aspect of industry 4.0 is the ability of machines, devices, and people to connect and communicate with each other via the Internet of Things. Inter-connectivity

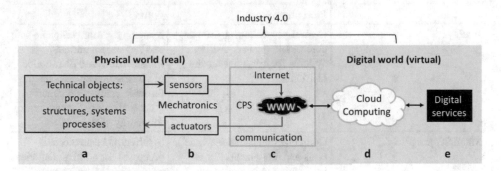

Fig. 5.2 The technological conception of Industry 4.0

allows operators to collect data and information from all points in the manufacturing process, thus aiding functionality. Industry 4.0 integrates processes in product development, manufacturing and service from suppliers to customers plus all key value chain partners.

With the help of cyber-physical systems that monitor physical processes, a virtual copy of the physical world can be designed. The technological conception of industry 4.0 and their essential parts are illustrated in Fig. 5.2.

a. Characterization of the physical elements of industry 4.0: products, structures, systems, processes
b. Combination of physical elements to mechatronic systems with sensors, actuators, computers and control modules
c. Extension of mechatronic systems to cyber-physical systems (CPS) and internet communication
d. Cloud computing with available to many users over the Internet
e. Digital services with webservers or mobile applications.

Typical application areas of the concept of cyber-physical systems in industry 4.0 are [4]:

Power engineering with CPS technologies for linking decentralized generation and distribution of electrical energy, in order to achieve an optimal, need-based and stable to ensure the functioning of energy networks (smart grids).
Transport technology with networking of vehicles ("Car-to-X") with each other or with the transport infrastructure via mobile radio. Cyber-physical systems will play a key role in future mobility, as they provide the basis for energy, battery and charging management.
Product and production systems, which are controlled by CPS via component, plant, factory and company borders are networked with each other. This enables rapid production according to individual customer requirements. Also, the production process within companies can be made more adaptive, evolutionary and cooperative through a network of worldwide self-organizing production units of different operators.

Medical technology systems, in which e.g. near-body sensors and medical information systems are connected via the Internet for remote monitoring. The acquisition of medical data by means of suitable sensor technology and their processing and evaluation in real time enables individual medical care.

References

1. Suh, S.C., Carbone, J.N., Eroglu, A.E.: Applied Cyber-Physical Systems. Springer, Heidelberg, Berlin (2014)
2. Office of Science and Technology Policy, Science and Technology for 21st Century Smart Systems. White Paper. www.CPS-OSTP-Response
3. Vogel-Heuser, B., Bauernhansl, T., ten Hompel, M.: Handbook Industry 4.0. Springer, Wiesbaden (2017) (in German)
4. Reinheimer, S. (Hrsg): Industrie 4.0—Herausforderungen, Konzepte und Praxisbeispiele. Springer, Wiesbaden (2017)

Systems Thinking in Health Technology   6

Health technology is defined by the World Health Organization as the "application of organized knowledge and skills in the form of devices, medicines, vaccines, procedures, and systems developed to solve a health problem and improve quality of lives".

Fundamental for the application of systems thinking in health technology are body functions and bio signals of the human body that can be recorded and therapeutically influenced [1].

A **biosignal** is any signal in living beings that can be continually measured and monitored, Fig. 6.1 gives an overview.

Biosignals characterize the medically relevant bodily functions of humans. They can be used with biosensor technology in systems diagnostics as functional variables (e.g. pressure, flow velocity, acoustic noise, temperature, electrical potentials) and as structural characteristics of humans (e.g. skeleton structure, organ dimensions, volumes, bone elasticity).

Biosignals are described—similar to signal functions in physics and technology—by signal shape, frequency, amplitude and the time of their occurrence. Their time response can be stationary, dynamic, periodic, discrete or stochastic. In medical technology applications, biosensors are combined with metrological components of signal processing and thus form a biological measurement chain, Fig. 6.2.

The technical systems of medical technology are medical devices: They are denoted in the EU *Directive on Medical Devices* (93/42/EEC) as follows: instruments, apparatus, substances or other objects which are used for the detection (diagnosis), prevention, surveillance (monitoring) and medical treatment (therapy) of human diseases or for the restoration of health (rehabilitation).

Medical devices are mechatronic systems. They consist of mechanics/electronics/IT components and work with sensor/processor/actuator functional elements. In their medical

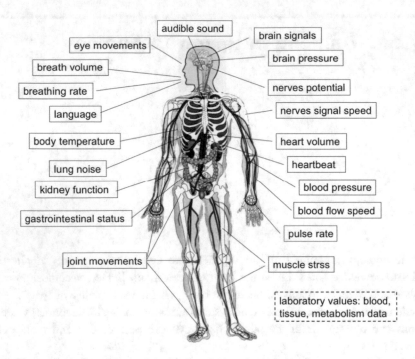

Fig. 6.1 Human body functions and the biosignals that characterize them

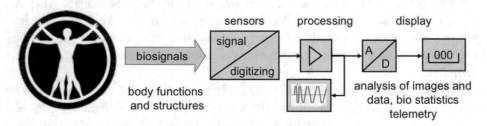

Fig. 6.2 Principle of a biosensor in a biological measurement chain

technology applications by the physician they interact with the human being as a patient—especially in diagnostics, monitoring and rehabilitation. In technological terms, medical equipment technology is extremely diverse and is divided into two major areas:

- Body sensors for body functions, e.g. electro physical measurements of heart or brain functions, ECG, blood pressure sensors,
- Imaging procedures, e.g. X-ray diagnostics, sonography (ultrasound diagnostics), computed tomography, magnetic resonance imaging.

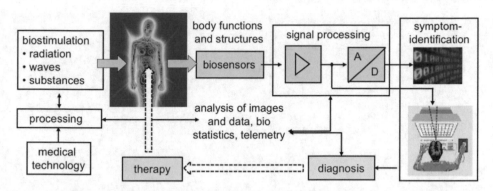

Fig. 6.3 Systems thinking in health technology: biostimulation and biosensors for diagnosis and therapy, supported by a variety of technological procedures and equipment

The application of medical technology by medical doctors is often associated with biostimulation, i.e. the application of stimuli, radiation, waves or medical substances to human test persons. This serves to improve the functionality of human sensory organs from the reaction to mechanical, acoustic, electrical, magnetic or visual stimuli. The determination of the resulting biosignals is carried out in a symptom analysis by biosensors using various methods:

- Sonography enables the analysis of ultrasound reflexes in the human body, the representation of movement sequences of heart valves and the measurement of blood flow velocity.
- X-ray absorption measurements allow the measurement of anatomical structures.
- By means of radioactively marked substances, metabolic phenomena and the transport speed, the accumulation site and the dynamics of precipitation processes can be determined.
- Ergonomic measurements can be used to determine the physical load capacity.

The principle of bioactuators and the interaction of biostimulation and biosensors is shown in a biological-mechatronic measuring chain in Fig. 6.3. The actual diagnostic findings are always provided by the responsible physician, supported by a variety of technological procedures and equipment.

Reference

1. Kramme, R. (ed.): Medical Technology. Springer, Heidelberg (2007) (in German)

Printed in the United States
by Baker & Taylor Publisher Services